A foundation course in chemistry

A FOUNDATION COURSE
IN CHEMISTRY

Bryan Slater
Education Officer, Cornwall County Council, formerly
Head of Chemistry, Bury Grammar School for Girls.

Jeff Thompson
Professor of Education, University of Bath, formerly Head
of Science Education, Department of Educational Studies,
University of Oxford

In conjunction with a panel of advisory teachers:

G. Evans, Matthew Arnold School, Cumnor
F. Lewis, Gosford Hill School, Kidlington
J. Piper, Peers School, Littlemore

M
Macmillan Education

First published 1982
Reprinted 1983, 1985

Published by
MACMILLAN EDUCATION LTD
Houndmills Basingstoke Hampshire RG21 2XS
and London
Companies and representatives throughout
the world

Printed in Hong Kong

British Library Cataloguing in Publication Data

Slater, Bryan
    A foundation course in chemistry.
    1. Chemistry
    I. Title   II. Thompson, Jeff
    540   QD33

ISBN 0 333 25515 1

CONTENTS

ACKNOWLEDGEMENTS

The author and publisher wish to acknowledge the following photograph sources:

COVER   John G. Darbey
Biophoto Associates p. 16
British Aluminium Co. Ltd. p. 12
British Tourist Authority p. 29
J. Allan Cash Limited p. 106
Central Electricity Generating Board p. 4
De Beers Consolidated Mines Limited p. 9
Esso Photograph p. 115
GKN p. 60
ICI Mond Div. Cheshire p. 14
Institute of Geological Sciences, London p. 10
L. and G. Fire appliance Co/Peter Bloomfield p. 95
NASA p. 43
Permutit-Boby p. 105
Popperfoto p. 6
Thermit Welding (GB) Limited p. 39

The publishers have made every effort to trace copyright holders, but if they have inadvertently overlooked any they will be pleased to make the necessary arrangements at the first opportunity.

PREFACE

This book has been written with the needs of all those who are beginning a course in chemistry in mind. Some students will have already gained a substantial amount of chemical knowledge, and will have acquired considerable skills, before embarking on a study of chemistry as a separate subject. However, there will also be those who will have little previous knowledge of the subject, and because of this the book contains no assumptions concerning previous chemical experience. The authors also realise that the expectations of the outcome of an initial study of chemistry will be different for different students, and in this respect the requirements of three particular groups of students have been recognised: those who will eventually specialise in science at G.C.E. Ordinary and Advanced level; those who will take chemistry for the Certificate in Secondary Education; and those for whom this book will constitute a terminal study of the subject. The theme of the book is the properties of some of the elements and the simple compound types which they form in reacting with one another. The study of the elements and compounds is, for the most part, treated at a concrete level appropriate for the majority of students beginning a study of chemistry. Students' understanding develops on the basis of their own first-hand experience. At the end of each major section, however, a brief explanation of what has been learned is suggested. Mastery of these theoretical models is not necessary for an understanding of subsequent sections, and the Course is complete in itself if they are omitted.

The book has been generated as a result of a collaborative venture between the authors and three teachers in different Oxfordshire comprehensive schools. The material is intended for use across a wide ability range, and because of this the problems of conceptual level and linguistic modes have been paramount. The material as originally generated was tested in the schools concerned, involving some one thousand children over a two year period, and modified in the final version as a result.

Details of apparatus and material required for the experimental work have been included as an appendix, as have been answers to multiple-choice and short-answer questions for each of the sections. The recommendation of the various Examination Boards and the Association for Science Education with respect to the use of nomenclature and units have been taken into account, although it has not been possible to be entirely rigorous in this respect.

The authors wish to thank all those who have contributed to the production of this book: the students who have commented freely, either implicitly or explicitly, during the trial of the material; the Advisory teachers for their expertise and patience in linking the work of the authors with the requirements of the schools; the Publishers who have been so generous in their encouragement, support and patience during the production of the material.

Bryan Slater
Jeff Thompson

Topic 1
DIFFERENT SORTS OF ELEMENT

A1.1 What is an element?

An element is a chemical, but a very special type of chemical. Elements are the simplest chemicals there are. Elements can go together to make complicated chemicals, but we can never make anything simpler than an element. All chemicals are either elements or substances more complicated than elements which have been made by putting different elements together. Elements are the *building blocks* for all chemicals.

We can use chemical reactions to make more complicated chemicals out of simpler ones, or to break up complicated chemicals. But we cannot go on breaking up complicated chemicals into simpler parts for ever. There comes a point where we cannot make chemicals any simpler. The chemicals which we cannot break up into simpler chemicals using chemical reactions are called elements.

An element is a chemical which cannot be split up into two or more simpler chemicals using a chemical reaction.

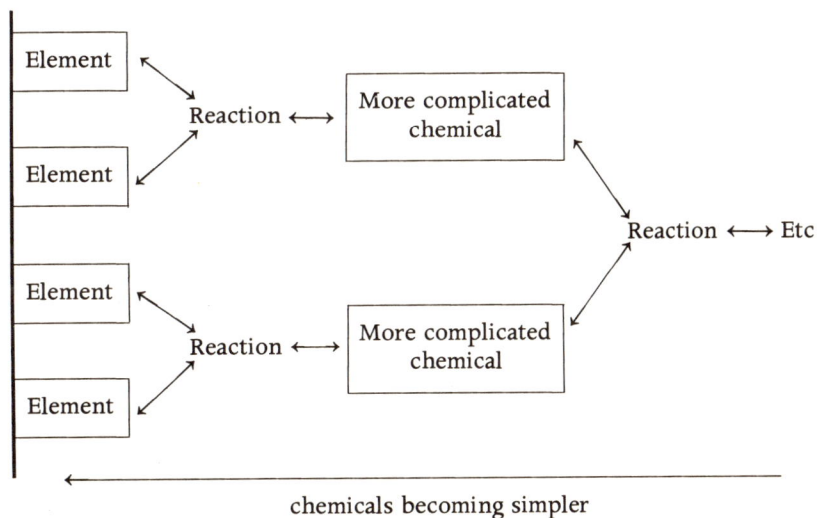

chemicals becoming simpler

We can often tell that we have succeeded in splitting up a chemical because we end up with something which weighs less than the thing we started with. We can, for instance, take the chemical water and break it up into two simpler chemicals, hydrogen and oxygen. The weight of hydrogen which we get is less than the weight of water which was used to form it. The difference between the two weights is equal to the weight of oxygen which was formed at the same time as the hydrogen. If we now take the hydrogen and try to split it up into simpler chemicals we will fail, however many chemical reactions we try. This is because hydrogen is an element and cannot be split up. Oxygen is also an element.

So water is one of those substances which is not an element itself, but which is made of different elements put together. These substances are called compounds.

A compound is a substance which is made of two or more elements combined together chemically.

Here is a list of some compounds and the elements which go together to make them:

Compound	Elements
Water	Hydrogen, oxygen
Table salt	Sodium, chlorine
Copper(II) sulphate	Copper, sulphur, oxygen
Nitric acid	Hydrogen, nitrogen, oxygen
Ammonia	Hydrogen, nitrogen

You will learn more about compounds in the later Sections of this book, but first you need to know more about elements.

A1.2 What do elements look like?

Your teacher will show you some elements.

Experiment 1

The following will be needed: samples of the elements sodium, potassium, magnesium, aluminium, iron, zinc, mercury, carbon, sulphur, bromine and iodine; gas jars containing chlorine and oxygen.

Your teacher will show you a number of elements. You may be allowed to examine magnesium, aluminium, iron, zinc, carbon, sulphur and iodine for yourself. Each element has its own name. You should write down the name of each element and next to its name record whether the element is a solid, a liquid or a gas, and what it looks like. If the element is a metal (your teacher will tell you if you are not sure) write this down next to the name.

You will end up with a table like the one below, although perhaps not in the same order.

Element	Metal or Non-metal	Solid, liquid or gas	Appearance
Sodium	Metal	Solid	Dull surface, shiny when cut. Kept under oil.
Potassium	Metal	Solid	Dull surface, shiny when cut. Kept under oil.
Magnesium	Metal	Solid	Dull surface, shiny when cleaned.
Aluminium	Metal	Solid	Quite shiny, especially when cleaned.
Iron	Metal	Solid	Dull surface, shiny when cleaned.
Zinc	Metal	Solid	Rather dull. Light grey in colour.
Mercury	Metal	Liquid	Very shiny. Reflecting surface.

Element	Metal or Non-metal	Solid, liquid or gas	Appearance
Carbon	Non-metal	Solid	Black powder.
Oxygen	Non-metal	Gas	Colourless.
Sulphur	Non-metal	Solid	Yellow powder.
Chlorine	Non-metal	Gas	Pale yellow/green.
Bromine	Non-metal	Liquid/gas	Deep red/brown.
Iodine	Non-metal	Solid	Purple/black crystals.

Elements which are not metals are called non-metals. Look at the table. Can you see any differences between metals and non-metals?

A1.3 How are metals different from non-metals?

You will carry out some tests on metals and on non-metals

Testing elements to see whether they conduct electricity

Perhaps there is a real difference between the elements which are metals and those which are non-metals. If we can show that this is so, it will be an important discovery to make. It will mean that elements can be grouped together into different types. To investigate this possibility it is necessary to carry out some tests on metals and non-metals to see how they behave.

Experiment 2

The following will be needed: power pack or battery (6 volts); bulb; connecting wires; crocodile clips or probes; samples of magnesium, iron, copper and aluminium; carbon rods and lumps of sulphur.

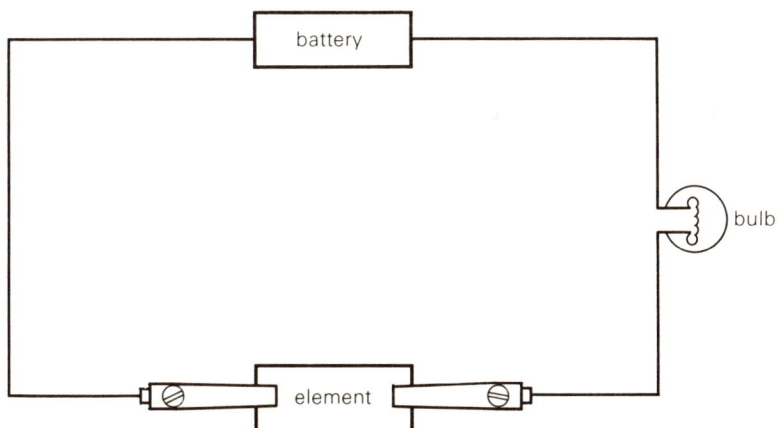

Your teacher will give you four metals (magnesium, aluminium, iron and copper) and two non-metals (carbon and sulphur) to test. Set up a circuit with a battery, a bulb and a gap for testing the element to see whether it conducts electricity. The ends of the gap should be two wires with crocodile clips which can be fastened to the element to test it. Test your circuit first, by touching the crocodile clips together. If the bulb lights up, the circuit is working and you can go on to test the elements in turn.

Test each element and write down (next to its name) whether or not it will conduct electricity. Write down what you think the difference is between metals and non-metals as shown by this experiment.

These pylons carry wires which conduct electricity

Testing the substances produced when elements burn

Experiment 3

The following will be needed: 2–3 cm length magnesium ribbon; a small piece of calcium; powdered sulphur; a test-tube containing carbon dioxide; tongs; emery paper; spatula; small beaker; broken porcelain; Universal Indicator paper; safety spectacles. The room must be well ventilated.

Wear your safety spectacles throughout this experiment.

Your teacher will give you three elements (magnesium, calcium and sulphur) and a test-tube containing carbon dioxide (the product of burning the element carbon). You will burn the three elements and test the substance produced each time. You will also test the carbon dioxide. In each case follow the instructions below.

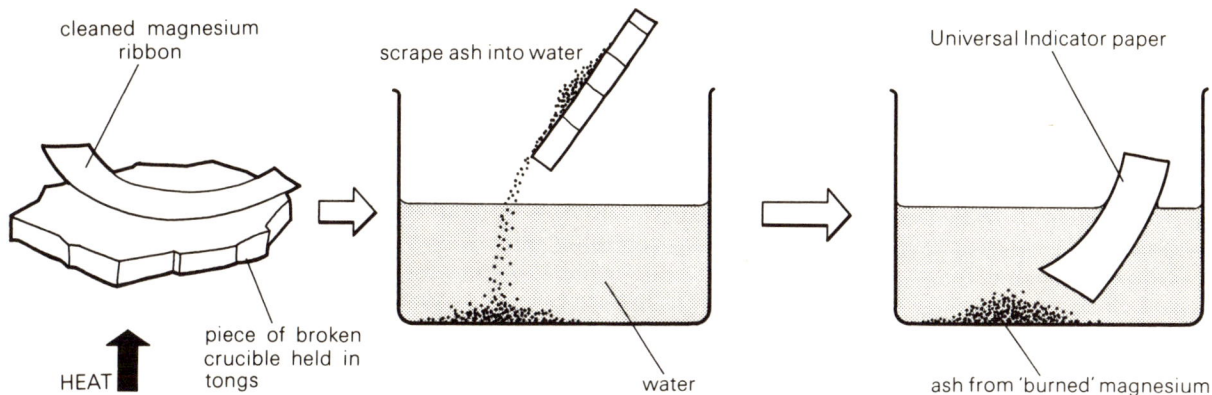

Magnesium. Clean the piece of magnesium ribbon until it is shiny all over. Put a little water (about 1 cm depth) in a beaker. Have this beaker ready on the bench.

Put the magnesium on a piece of broken porcelain and hold the porcelain carefully in a pair of tongs. Hold it in a roaring Bunsen burner flame until the magnesium begins to burn. *Do not look directly at the burning magnesium because the light will hurt your eyes.*

When the magnesium has finished burning, scrape the ash into the beaker of water with a spatula. Use the spatula to stir the water. Then test the solution with Universal Indicator paper. Write down whether the solution is acidic, alkaline or neutral.

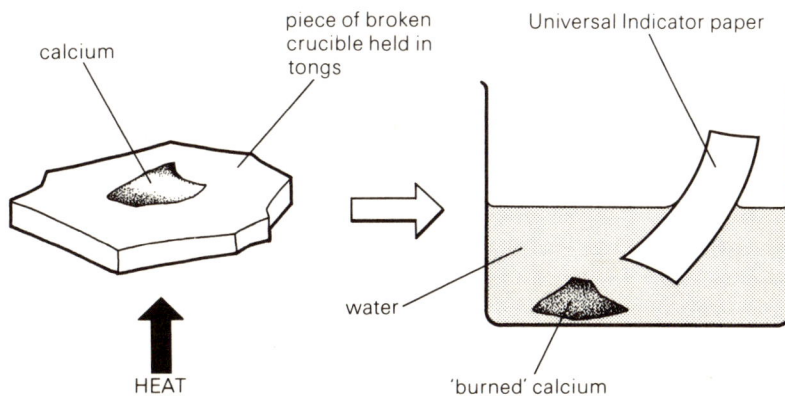

Calcium. Wash out the beaker and refill to 1 cm depth with water. Strongly heat a small piece of calcium on a piece of broken porcelain for about five minutes. At the end of this time, put the calcium in the beaker and stir with a spatula. Test the solution with Universal Indicator paper and write down whether it is acidic, alkaline or neutral.

damp Universal
Indicator paper

sulphur powder

broken crucible

HEAT

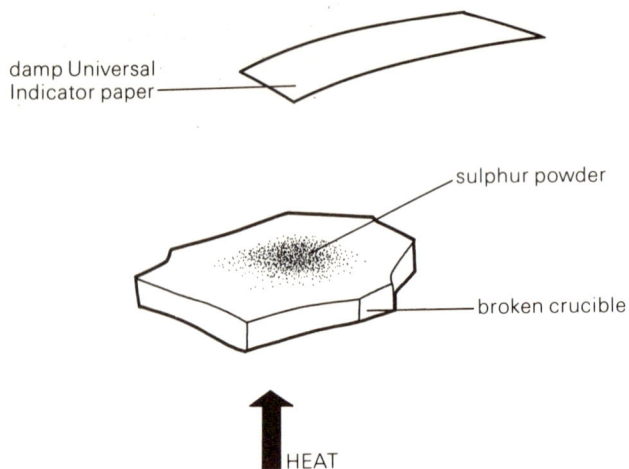

Powdered sulphur. Place a small spatula measure of powdered sulphur on a piece of broken porcelain. Hold this in a pair of tongs and heat it in a roaring Bunsen burner flame. From time to time take the broken porcelain out of the flame and test the gas given off with *damp* Universal Indicator paper. This gas is the product of burning sulphur in air. It is very unpleasant if breathed in in quantity, so take care not to get too close to the heated sulphur. Write down whether the gas is acidic, alkaline or neutral.

When a volcano erupts very hot elements such as sulphur and iron burn in air

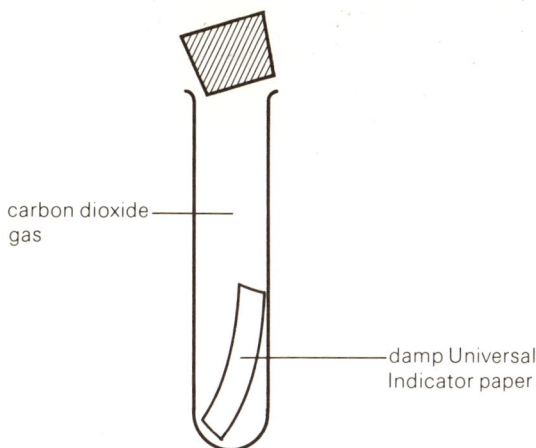

carbon dioxide gas

damp Universal Indicator paper

Carbon dioxide. It is difficult to make the product of burning carbon by heating this element in the air as was possible for the other elements. You will be given a test-tube containing the gas carbon dioxide (or if time is short your teacher will show you this test). Carbon dioxide is the product of burning carbon. Take a piece of damp Universal Indicator paper, remove the bung from the test-tube and test the gas. Write down whether it is acidic, alkaline or neutral.

Write down the difference between metals and non-metals which this experiment has shown.

Topic summary and learning

1   All substances are made up of different elements combined together chemically, or are elements themselves. Elements are the simplest chemicals which exist. Substances which are not elements are called compounds.

2   Elements can be divided into two types – metals and non-metals. We can say the following about each type of element:

Metals

They are usually solids. Mercury is the only metal which is a liquid at room temperature. Metals are never gases under normal conditions.

Their surfaces are usually shiny (if clean).

They can be very strong materials physically.

They all conduct electricity.

The substance produced when a metal burns in air is called a *base*. Many bases are soluble in water and if they are they form alkalis. An alkali is the solution formed when a soluble base dissolves. When metals burn in air they therefore often produce substances which dissolve in water giving solutions which are alkaline.

Non-metals

They can be solids, liquids or gases under normal conditions.

If they are solids they are physically weak, and are not shiny.

They do not usually conduct electricity. Carbon (in the form of graphite) is the only common non-metal which does conduct electricity.

The substance produced when a non-metal burns in air forms an acid when it dissolves in water.

Learn the information contained in the Topic Summary.

You should learn the definition of what an element is and what a compound is.

You should know the general characteristics which make metals and non-metals different from each other.

You should remember the names of the metals and non-metals which you have tested, and how they behave.

Questions

1   Explain *in your own words* what an element is.

2   Suppose you had to choose *one* test to carry out on elements to decide whether they were metals or non-metals. Which test would you choose, and why?

3   Here are some facts about some more elements. For each element decide whether it is a metal or a non-metal, and give reasons for the decision you make.

Nitrogen. It is a colourless gas. Like carbon, it is difficult to burn nitrogen in air. However, we can obtain the same substance that would be produced by burning if we pass a spark through a mixture of nitrogen and oxygen. This substance dissolves in water to form an acid.

Lithium. It is a solid which has to be kept under oil. It has a shiny surface when it is cut. The substance formed when it burns in air dissolves in water to form an alkali.

Phosphorus. It is a white (or red) solid. It is not physically strong. It burns in air producing a substance which dissolves in water to give a solution which turns Universal Indicator paper red.

Calcium. It is a solid. Its surface is shiny when it is cut. The substance formed when it burns in air gives an alkali in water.

4   For each of the following elements, give one further property it is likely to have.

Copper. It is a metal. The substance produced when it burns in air is insoluble in water.

Rubidium. It is a metal. It has to be kept under oil.

Oxygen. It is a non-metal. It does not burn (it would have to be reacting with itself to do that!).

Topic 2
GETTING ELEMENTS

A2.1 Where do we find elements?

The short answer to this question is that we find elements everywhere, since all substances contain elements. However, we often want to get hold of the elements *on their own*, and not combined with other elements in compounds.

A few elements *are* found on their own. Copper, silver and gold can be dug out of the ground in some parts of the world. We have only to separate them from the impurities we dig out of the ground with them. Also, air is a mixture of elements. If we separate the gases in the air we can obtain the elements oxygen, nitrogen, neon, argon, krypton and helium on their own.

The element carbon is dug out of the ground in the form of diamonds

Most elements, however, are contained in compounds. To get them on their own we have to break up the compounds which contain them. When we set about doing this, we must choose carefully the compounds which are to be broken up. For instance, human beings are made of the elements carbon, nitrogen, hydrogen and oxygen, together with a few metals. Nobody would dream of trying to get these elements on their own starting from human beings! We can get the same elements much more easily (and less painfully!) from coal, water, the air and some rocks.

A2.2 Practical, economic and environmental considerations

The compounds we choose to provide us with elements are known as *ores*. In deciding on the most useful ore to use for a particular element, a chemist has to consider three important things. Firstly, he has to take into account how easy it is to obtain the ore in the first place. Then he has to consider how much it will cost to get the element from the ore. *The ore that is chosen is almost always the one that will allow us to get the element for the lowest total cost.* So we choose ores which are plentiful, and from which we can easily obtain the elements we want. It is also important not to forget the damage which we might cause to the surroundings (the amount of "environmental pollution" we create) when we dig up the ore and when we carry out the process of getting the element we want from it.

Rocks, the sea and the air are the most convenient sources of

Haematite, a common ore of iron

the majority of the elements we want. Among the most common elements obtained from rocks are calcium, carbon, silicon, copper, aluminium, iron, lead and zinc. The sea provides us with massive sources of sodium, magnesium, chlorine, bromine, hydrogen and oxygen.

But how do we get elements from their ores?

A2.3 Getting elements from compounds

You will carry out experiments to get elements from compounds which contain them.

Using electricity to produce metals

We shall look at three methods by which elements can be produced from compounds:
a) using electricity to split up solutions of compounds
b) using other elements to split up solid compounds
c) using other elements to split up solutions of compounds.

Experiment 4

The following will be needed: power pack or batteries (12 volts); connecting wires; crocodile clips; old carbon electrodes; small beaker; copper(II) sulphate solution; emery paper.

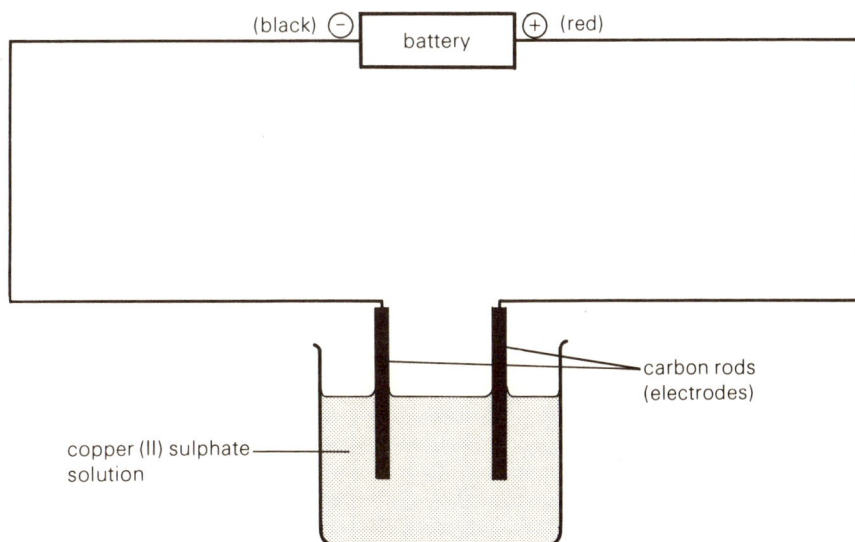

Clean any deposit from the electrodes using emery paper and assemble the circuit as shown. The electrodes must not be allowed to touch, so use electrode holders if these are available. If not, place an empty test-tube between the electrodes to keep them apart. Allow the experiment to run for about 10 minutes. At the end of this time switch off and examine each electrode. You will see that copper has been deposited on the one that was attached to the negative terminal of the battery.

Where has this copper come from? Write down what you think this experiment has shown.

Before returning the equipment, gently scrape off the deposit of copper from the electrode using firstly a spatula and then emery paper.

Aluminium is obtained by electrolysis of its ore, bauxite. This shows one of the cells in which the reaction takes place

Electricity is used to produce copper from its ore in industry. It is also used to obtain sodium, magnesium and aluminium. The cost of carrying out these processes depends on the cost of the electricity which is used, and so these elements tend to be produced in parts of the world where electricity is cheap.

Using electricity to produce non-metals

Experiment 5

The following will be needed: power pack or batteries (12 volts); connecting wires; container with two fixed carbon electrodes; very strong solution of common salt; test-tubes, litmus paper; splint; bungs for test-tubes.

Assemble the apparatus as shown. The test-tubes are filled with the solution of salt (often called "brine") and inverted over the carbon electrodes. When the circuit is connected, bubbles should be produced at both electrodes. Make sure the voltage is high enough for the gases to be given off fairly quickly.

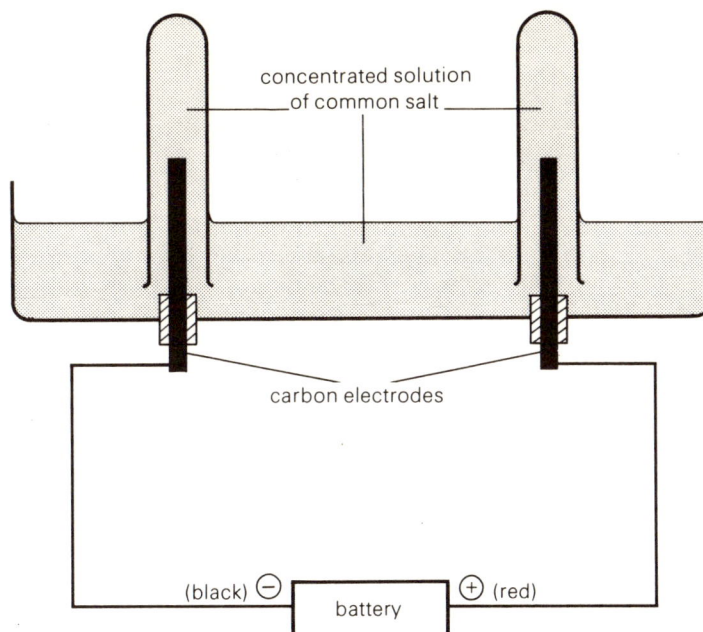

concentrated solution
of common salt

carbon electrodes

(black) ⊖   battery   ⊕ (red)

Allow the experiment to run until both test-tubes are full of gas. Then remove the test-tubes and place a bung in each. Remember which gas was produced at which electrode.

Examine each gas to see if it has a colour. Drop a piece of damp red (or blue) litmus paper into each, and quickly replace the bung. Observe what happens to the litmus paper. Remove the bung again and quickly test each gas with a glowing splint. Write down your results for each gas. Can you identify the gases?

The gas produced at the electrode connected to the negative terminal of the battery is hydrogen. It is colourless, has no effect on damp litmus paper, and explodes with a characteristic noise. The other gas (produced at the electrode connected to the positive terminal) is chlorine. It is yellow/green, removes the colour from (bleaches) litmus paper, and does not burn. So this experiment has shown that non-metals (chlorine here) can be produced from compounds using electricity, in just the same way that metals can. Chlorine is needed in large quantities in industry, and it is produced using electricity.

Common salt is made up of the elements sodium (a metal) and chlorine (a non-metal). Where do you think the hydrogen in this experiment came from? You will learn later why sodium is not produced in this experiment as might be expected.

If you look back over the last two experiments, you will see that when electricity is used to split compounds up into their elements, metals and hydrogen are produced at the negative electrode and non-metals are produced at the positive electrode. So we can use electricity to produce two elements at the same

Chlorine is obtained industrially by the electrolysis of concentrated sodium chloride solution

time. If we arrange matters so that both the elements are elements which we want, we can get a metal and a non-metal for a given amount of electricity. Therefore when metals are produced using electricity, ores are chosen which result in chlorine being produced at the same time at the other electrode.

When we use electricity to split a compound up into its elements, the process is called *electrolysis*. You will learn more about electrolysis in a short while.

Using other elements to split up solid compounds

Experiment 6

The following will be needed: charcoal block; blow pipe; lead oxide (litharge); safety spectacles.

Wear your safety spectacles throughout this experiment.

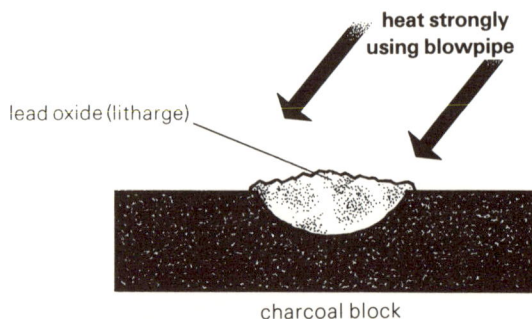

Put a small amount of lead oxide into a hole in a charcoal block. (If there is not a hole there already, your teacher will show you how to make one using a coin.) Your teacher will show you how to heat a small area of the charcoal using a blowpipe. Heat the area around the lead oxide very strongly indeed. Do not blow too hard or you will probably blow out all the lead oxide. (You can moisten the lead oxide with a drop of water to make it stay on the block).

Note any change in the colour of the oxide as you heat the block. After some time, beads of lead will appear round the edge of the oxide and sometimes these will form a large bead of the silvery metal.

Write down what you think the charcoal block (the element carbon) has done to the lead oxide.

Lead oxide is a compound of two elements – lead and oxygen. By heating lead oxide with carbon, lead on its own is produced. The lead and the carbon were "in competition" for the oxygen, and the carbon won. It "grabbed" the oxygen, leaving lead on its own. A compound of carbon and oxygen, carbon dioxide, was also produced. What happened to the carbon dioxide produced in your experiment? Carbon is good at "grabbing" the other element from compounds containing elements we want. We make carbon react with ores to produce the elements zinc, iron and lead.

Using other elements to split up solutions of compounds

The following experiment is best demonstrated by the teacher, using an overhead projector if one is available.

Experiment 7

The following will be needed: silver nitrate solution; small strip of magnesium ribbon; glass petri dish; overhead projector (if available).

Place a small amount of silver nitrate solution into the petri dish and put a small piece of magnesium ribbon in the middle of the solution. If the dish is placed on an overhead projector, crystals can be seen growing out from the magnesium on the projected image, in the space of about 20 minutes.

What do you think the crystals are? Write down what the magnesium has done to the silver nitrate solution.

In this experiment silver has been "displaced" from the solution of the compound silver nitrate by the magnesium. You will see more examples of these "displacement" reactions later on in the section on "reactivity". This experiment could also be

Crystals of silver obtained from silver nitrate solution

called a "crystallisation" since crystals of silver are formed. The less reactive element is the one that is left at the end of a displacement reaction. Here, we have shown that silver is less reactive than magnesium. In the previous experiment we saw that carbon is more reactive than lead.

We could use "displacement" to produce elements like iron. For instance, we could take a solution of a compound of iron and displace the iron with a reactive metal like magnesium. However, this would be very expensive, since magnesium is expensive to buy. So a cheaper method is chosen. We obtain iron from its ore by heating it strongly with carbon (in the form of coke) in a Blast Furnace. This process is carried out on a very large scale in industry.

Topic summary and learning

1   A few elements (copper, silver and gold) are sometimes found on their own. Most elements, however, are found combined with other elements in compounds.
2   The air is not a compound, but it is a mixture of the elements oxygen, nitrogen, neon, argon, krypton and helium. These can easily be separated.
3   We obtain the elements calcium, carbon, silicon, copper, aluminium, iron, lead and zinc from rocks.
4   We obtain the elements sodium, magnesium, chlorine, bromine, hydrogen and oxygen from the sea.
5   We can obtain elements from solutions of their compounds by electrolysis. Metals are produced at negative electrodes and

non-metals at positive electrodes during electrolysis. The elements copper, sodium, magnesium and aluminium (metals) and chlorine (non-metal) are amongst those obtained by electrolysis.

6   Another way of obtaining an element from a compound containing it is to heat the compound with another element such as carbon. Carbon is used to produce the elements zinc, iron and lead from their ores.

7   We can also use a very reactive metal like magnesium to obtain elements from solutions of their compounds, but this is usually too expensive. We have to consider the cost of obtaining the compounds containing the element we want, and the cost of getting the element out of this compound before we decide how to go about getting a particular element. We also have to take environmental factors into account in making this decision.

Learn the information contained in the Topic Summary.

You should remember what electrolysis is, and what happens when electrolysis is carried out.

You should know where each of the following elements is found, and the process by which it is obtained from the compound containing it: copper, sodium, magnesium, aluminium, chlorine, zinc, iron and lead.

You should remember how to identify the gas chlorine.

Questions

1   What element would you expect to be produced at the *negative* electrode during the electrolysis of each of the following *molten* compounds: sodium chloride, magnesium chloride, aluminium chloride?

2   What element would you expect to be produced at the *positive* electrode in each of the electrolysis experiments in question 1?

3   Which of the following elements is the most reactive?

A.   zinc

B.   lead

C.   carbon

D.   iron.

4   Which of the following elements is the most reactive?

A.   iron

B.   magnesium

C.   silver.

5   The metal aluminium could be produced from a number of different materials. Use the information in the table on the next page to decide which method of obtaining aluminium is used in industry and give reasons for your choice.

| Method | Likely cost/unit weight of carbon | | Associated pollution |
	Source of aluminium	Extraction process	
A	Clay (£10)	Reaction with another element (£2000)	Large areas of landscape destroyed
B	Bauxite (£200)	Electrolysis (£200)	Acceptable
C	Precious stones (e.g. rubies) (£3000)	Reaction with another element (£2000)	Acceptable

6   The non-metal carbon could also be produced in a number of different ways. Using the following information, decide which method is used in industry and give reasons for your choice:

| Method | Likely cost/unit weight of carbon | | Associated pollution |
	Source of carbon	Extraction process	
A	Diamonds (£5000)	Burning to give carbon dioxide (£500)	Acceptable
B	Trees (£50)	Burning giving charcoal (£5)	Large-scale destruction of forests
C	Coal (£150)	Burning giving coke (£5)	Acceptable
D	Carbon dioxide (£100)	Reaction with another element (£2000)	Acceptable

7   The cheapest method for producing the element chlorine is by the electrolysis of concentrated sea-water, provided that a cheap supply of electricity is available. Suppose you had to decide between the following countries as places for manufacturing chlorine. Which country would you choose, and why?

Country	Distance from sea (km)	Cost of electricity (/unit)	Distance to customer (km)
A	5000	£5	2000
B	20	£50	2000
C	3000	£100	3000
D	20	£5	200
E	5000	£50	300

8   A lot of money and effort is spent on obtaining elements from their different sources in nature. But why are elements so important to us? Make a list of common elements and their uses. You should be able to think of everyday uses for the elements hydrogen, helium, carbon, oxygen, magnesium, aluminium, phosphorus, chlorine, chromium, iron, copper, silver, tin, iodine, gold, mercury and lead, and possibly many others.

Topic 3
LABELLING ELEMENTS

A3.1 Names and symbols of the elements

Each different element has a different name. But to save having to write these names out in full each time, we have a *symbol* for each element as well as a name. So we have a kind of "chemical shorthand".

The symbol for an element is either one letter or two. The first letter is always a capital letter. If there is a second letter it is not a capital.

You will get used to this chemical shorthand with practice, but first you need to know the symbols for some elements.

Element number	Name	Symbol	Element number	Name	Symbol
1	Hydrogen	H	11	Sodium	Na
2	Helium	He	12	Magnesium	Mg
3	Lithium	Li	13	Aluminium	Al
4	Beryllium	Be	14	Silicon	Si
5	Boron	B	15	Phosphorus	P
6	Carbon	C	16	Sulphur	S
7	Nitrogen	N	17	Chlorine	Cl
8	Oxygen	O	18	Argon	Ar
9	Fluorine	F	19	Potassium	K
10	Neon	Ne	20	Calcium	Ca

Some other elements and their symbols are:

Iron   Fe   Uranium U

Silver Ag   Copper   Cu

Lead   Pb   Mercury Hg

Gold   Au   Nickel   Ni

Zinc   Zn

This next exercise will help you to learn the names and symbols of the elements.

Chemical crossword

In the crossword on the next page, all the letters should be written as capitals. (But remember that not all the letters in chemical symbols are capitals.)

19

Your teacher will give you a copy of this crossword.

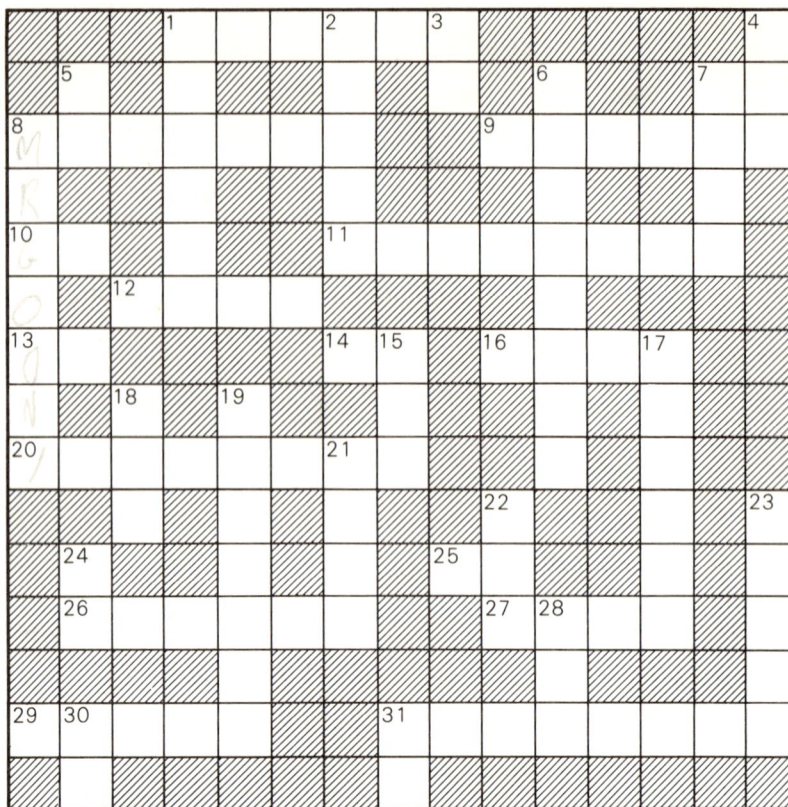

Clues across

1   A non-metal. It's usually black.
7   The letters we use for sodium.
8   A non-metal, with symbol S.
9   If you clean magnesium it does this.
10 The letters we use for lithium.
11 The name of the element with symbol N.
12 The colour of zinc.
13 The symbol for 6 down.
14 Does sulphur conduct electricity?
16 A metal, with symbol Zn.
20 Carbon (in the form of graphite) is the only one which conducts electricity.
25 Symbol for calcium.
26 Mercury is one.
27 You do this to make things dissolve.
29 Metals have this kind of surface.
31 Its symbol is H.

Clues down

1   A metal with symbol 5 down.
2   Its symbol is B.
3   The letters we use for 7 down.
4   Oxygen is one.
5   The symbol for 1 down.
6   The element with symbol Cl.
7   The name of a gaseous element whose symbol is 3 down.
8   Its symbol is Si.
15 We keep sodium in it.
17 Oxygen does not have any.
18 An element contains this number of different chemicals.
19 The only liquid metal.
21 You can make one by burning sulphur in air.
22 Chlorine is one at room temperature.
23 The colour of 6 down.
24 The letters we use for aluminium.
28 You get this from crude oil, and often find it on roads.
30 The letters we use for helium.
31 The letters we use for 19 down.

There are over 100 different elements. Learning all about them one by one would not only be difficult but it would also take a long time. Fortunately, chemistry is not as difficult as this. In most chemical laboratories you will find a table of the chemical elements; it is called a *Periodic Table*. This table is a shorthand representation of a lot of what we know about chemistry.

The elements are arranged in a particular way in the Table. The position of the element tells us a lot about the way that element behaves chemically.

Here are the first twenty elements arranged in the same way as they are in a full Periodic Table:

$\boxed{H}$						$\boxed{H}$	He
Li	Be	B	C	N	O	F	Ne
Na	Mg	Al	Si	P	S	Cl	Ar
K	Ca						

Hydrogen is the only element which does not really have a "best position" in the Table. Every other element only fits in one place, but hydrogen can fit into either of the two positions shown.

The full form of the Periodic Table (showing all the elements) is given on p. 22. It has the following shape:

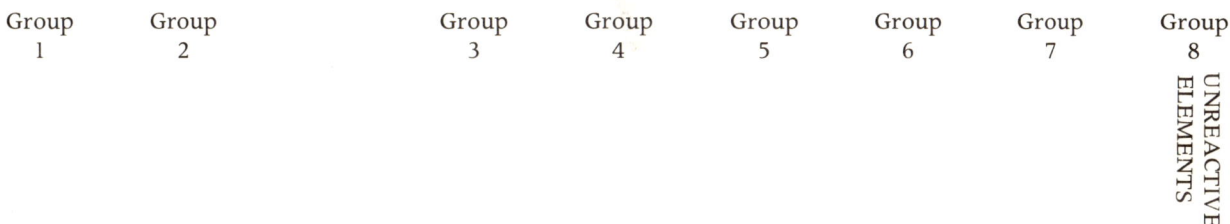

Group 1	Group 2		Group 3	Group 4	Group 5	Group 6	Group 7	Group 8
								UNREACTIVE ELEMENTS

The way we look at the Periodic Table is this. We call the vertical columns *groups* of elements. *Elements which are in the same group (vertical column) are similar to each other chemically.* There are eight such groups, called groups 1–8, starting with the group at the extreme left-hand side of the Table (group 1).

Since elements which are in the same group are similar to each other, we don't have to learn about all the elements individually. For instance, suppose you wanted to know something about the element beryllium (symbol Be). It is in the same group in the Periodic Table as magnesium (Mg) and calcium (Ca). So it behaves a lot like magnesium and calcium.

Name an element which you would expect to behave chemically in a similar way to chlorine.

Which elements behave chemically like neon?

Which elements behave chemically in a similar way to sodium?

The periodic table

──────────────── 8 groups of elements ────────────────

Gp.1	Gp.2												Gp.3	Gp.4	Gp.5	Gp.6	Gp.7	Gp.8
				Metals			Non-metals										H	He
H																		
Li	Be	The Transition Elements (metals)											B	C	N	O	F	Ne
Na	Mg	(A block of thirty elements)											Al	Si	P	S	Cl	Ar
K	Ca	Sc	Ti	V	Cr	Mn	Fe	Co	Ni	Cu	Zn		Ga	Ge	As	Se	Br	Kr
Rb	Sr	Y	Zr	Nb	Mo	Tc	Ru	Rh	Pd	Ag	Cd		In	Sn	Sb	Te	I	Xe
Cs	Ba	La	Hf	Ta	W	Re	Os	Ir	Pt	Au	Hg		Tl	Pb	Bi	Po	At	Rn

5 rows of elements

Metals │ Non-metals

Groups of metals are found towards the left-hand side of the Periodic Table, and groups of non-metals towards the right-hand side. Some groups (groups 3, 4, 5 and 6) contain elements which are metals and elements which are non-metals. For instance:

Group 6

O ⎫
S ⎬ are non-metals
Se ⎭
Te
Po   is a metal.

So, even though these elements are in some way similar, they are not identical. There must be a gradual change in the way they behave as we move further and further down the group.

We can now study the way chemical behaviour varies within groups of similar chemical elements by looking at two of the most interesting groups – group 1 (a group of metals) and group 7 (a group of non-metals).

A3.2 The group 1 elements: the "alkali metals"

Li   Lithium

Na   Sodium

K   Potassium

Rb   Rubidium

Cs   Caesium

> Your teacher will show you some reactions of the alkali metals. You may be allowed to carry out some of the reactions of lithium yourself.

You already know from Topic A1 that there is some similarity between these elements. First of all, they are all metals. You have seen that lithium, sodium and potassium are all kept in bottles of oil to stop air or moisture reaching them. We have to do this because these elements are all very reactive. Fortunately, they do not react with oil. The other members of the group are rubidium and caesium. They are so reactive we do not usually keep them in schools. However, we may be able to discover what rubidium and caesium are like by finding out about the behaviour of the other members of the group.

The following experiment must be demonstrated by the teacher.

Experiment 8

The following will be needed: samples of lithium, sodium and potassium; 3 gas-jars containing chlorine; 3 deflagrating spoons; 2 beakers containing small quantities of water; Universal Indicator paper.

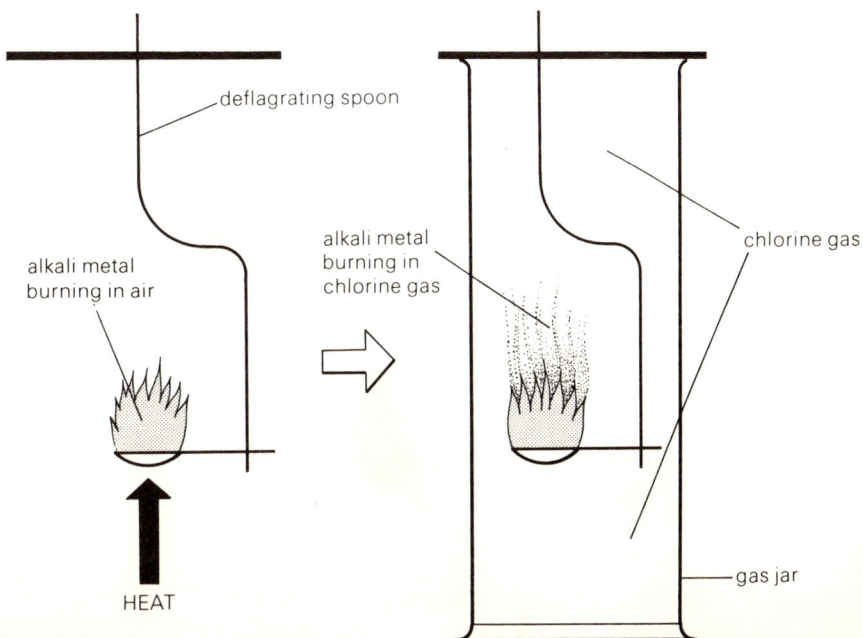

Obtain a small piece of lithium by cutting a larger piece on a tile with a knife. Place this on a deflagrating spoon and ignite it in the air. While it is still burning, plunge it into the chlorine gas. Repeat using sodium and potassium instead of lithium.

Take a small piece of sodium and add it to water in a beaker. Add Universal Indicator paper to the solution produced. Repeat using potassium instead of sodium.

While the teacher demonstrates these experiments, you should notice and write down the following for each of the alkali metals lithium, sodium and potassium:

1   How is the metal handled?
2   How hard is the metal to cut?
3   What does the metal look like when it has just been cut?
4   Does the element burn in air? How easily does it start burning?
5   Describe what happens when the element burns in chlorine.
For sodium and potassium:
6   What happens when the metal is added to water?
7   What can we say about the solution which is produced when this happens?

Pupils can carry out the following experiment.

Experiment 9

The following will be needed: 2 small pieces of lithium; broken porcelain; tongs; beakers; Universal Indicator paper.

Wear your safety spectacles throughout this experiment.
Do not touch the lithium with your hands.

Take a small piece of lithium which your teacher has cut for you, and place it on a small piece of broken porcelain. Hold this in a pair of tongs and heat strongly until the lithium burns. Compare what happens with the burning of sodium and potassium in air just carried out by your teacher.

Take a second small piece of lithium and add it to a small amount of water in the bottom of a beaker. Note what happens. When the reaction has finished, add Universal Indicator paper to the solution. Note the result.

Using the results of your teacher's and your own experiments, you should be able to draw up a table like the one on the next page. Now you can see why we have to keep these elements out of the way of air and water!

Test	Lithium	Sodium	Potassium
Hardness	Hard to cut (about the same as cheese).	Fairly difficult to cut (about the same as cold butter).	Fairly easy to cut (about the same as soft butter).
Reaction with air	Burns after strong heating for a few minutes.	Burns after strong heating for about half a minute.	Burns at once with strong heating.
Reaction with chlorine	Burns. Produces clouds of smoke.	Burns quite easily. Produces clouds of smoke.	Burns very easily. Produces clouds of smoke.
Reaction with water	Moves about on the surface. Solution is alkaline.	Moves about on the surface. Sometimes the gas produced catches fire. Solution is alkaline.	Moves about on the surface. The gas produced always catches fire. Solution is alkaline.

You should be able to see that these elements are similar in the way they behave, and also that they look alike. So the Periodic Table certainly works for this group of elements at least. Why do you think this group is called Group 1? The other name for this group (the "family name") is the *alkali metals*. Can you say why?

Have a closer look at what you know about these elements. They are similar, but are they the same? Going down the group, do they become more reactive or less reactive? We could summarise what we have found as follows:

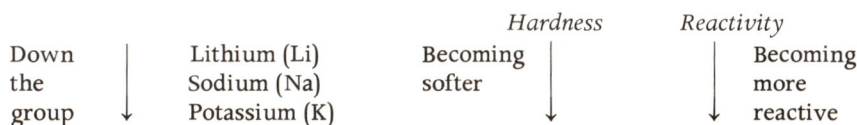

			Hardness	Reactivity
Down the group ↓	Lithium (Li) Sodium (Na) Potassium (K)		Becoming softer ↓	Becoming more reactive ↓

A3.3 The group 7 elements: the "halogens"

F   Fluorine
Cl Chlorine
Br Bromine
I   Iodine

Your teacher will show you some reactions of chlorine, bromine and iodine. You will carry out some reactions of iodine yourself.

In Topic A1 you saw examples of some of these elements. They are all non-metals. Chlorine is a yellow/green gas, bromine is a red liquid/gas and iodine is a black/purple solid. Fluorine is a faintly yellow gas.

Like the alkali metals, the group 7 elements are very reactive. Fluorine is so reactive that it even attacks glass apparatus, so we have to keep it in special containers. Because of this, we never see fluorine in school laboratories. However, we will be able to discover a lot about the way fluorine behaves by finding out how the other group 7 elements react.

The other name for this group (the "family name") is the *halogens*. This means "salt makers". Can you say why these elements have been given this name?

The following experiment must be demonstrated by the teacher in a fume cupboard.

Experiment 10

The following will be needed: chlorine generator; bromine; iron wool; test-tube with hole at the end; 2 boiling tubes; 1 test-tube and bung; Universal Indicator solution.

Pass chlorine gas into some water in a boiling tube. Add Universal Indicator solution.

Add a drop of bromine to a test-tube half filled with water. Insert the bung and shake vigorously. Add Universal Indicator solution.

Pass chlorine over heated iron wool as shown below, allowing pupils to observe whether a reaction has taken place.

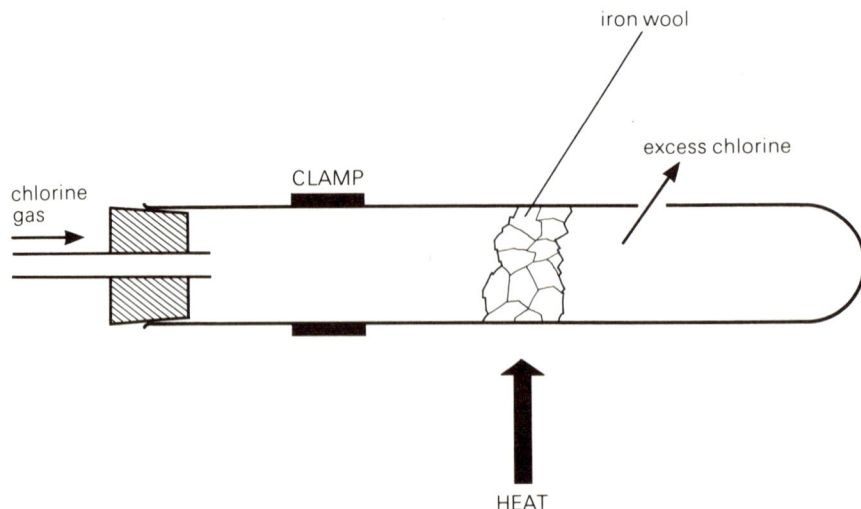

Pass bromine vapour over heated iron wool in a boiling tube as shown on p. 27, again allowing pupils to observe any reaction. The Bunsen flame need not be applied directly to the bromine. Continue heating the iron wool and sufficient heat will be transmitted to the bromine to vaporise it.

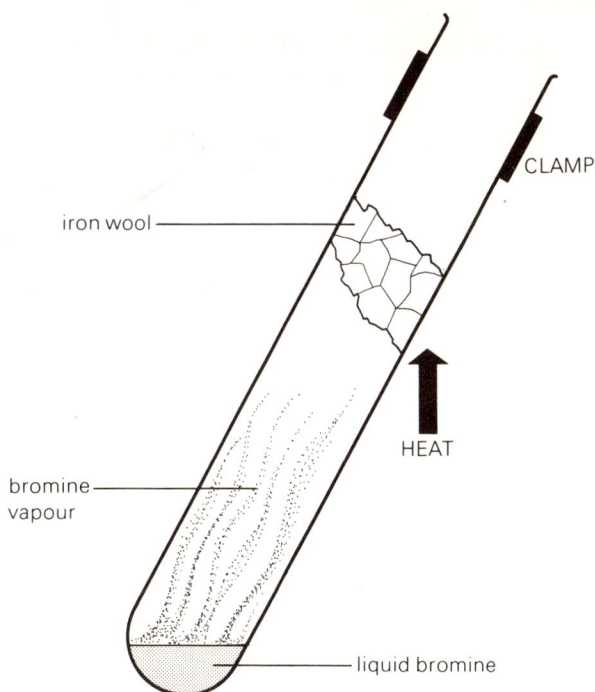

While the teacher demonstrates these experiments, you should make the following observations. Note down what you see.

1 Does chlorine appear to dissolve? How can you tell? Does it dissolve easily?

2 Is the solution of chlorine in water acidic, alkaline or neutral?

3 Does bromine appear to dissolve? How can you tell? Does it dissolve easily?

4 Does chlorine react with iron? How can you tell? How easily does it react?

5 Does bromine react with iron? How can you tell? How easily does it react?

Pupils can carry out the following experiment.

Experiment 11

The following will be needed: 2 crystals of iodine; iron wool; test-tube and bung; boiling tube; Universal Indicator paper.

Place a small crystal of iodine in a test-tube. Add cold water, insert the bung and shake. Is the iodine soluble? Remove the bung and warm the water carefully. Will the iodine crystal dissolve now? If it does, test the solution with Universal Indicator paper. Note the result.

Pass iodine vapour over *heated* iron wool, using the same apparatus that your teacher used for bromine. *Use one crystal of iodine only.* You may need to heat the iodine crystal *gently* to make it vaporise. Does the iodine react with iron? How easily does it react?

Using the results of your teacher's and your own experiments, you should be able to draw up a table containing the following information.

Test	*Chlorine*	*Bromine*	*Iodine*
Reaction with water	Dissolves. Forms an acid.	Dissolves. Forms a weak acid.	Dissolves with difficulty. Forms a neutral solution.
Reaction with iron	Reacts easily. Iron glows red.	Reacts. Iron has to be heated.	Reacts. Iron has to be heated.

So the halogens are similar, but not identical in the way they behave. Chlorine reacts more easily with both water and iron than does bromine, and bromine in turn is more reactive than iodine. As we go *down* the halogen group in the Periodic Table, the elements become *less* reactive. So the *reactivity trend* is in the opposite direction from that for the alkali metals. The elements also become *harder* as we pass down the group.

		State	*Reactivity*
Down	Chlorine (Cl)	Gas	Becoming
the	Bromine (Br)	Liquid	less
group	Iodine (I)	Solid	reactive

A3.4 The "transition" elements

One part of the Periodic Table has to be treated differently from the rest. Look at the full form of the Table (p. 22). You will see that in the centre of the Table there is a block of elements separated from the rest. These elements do not belong to vertical groups, and they are called "transition" elements. *The transition elements are the only ones that do not belong to vertical groups.*

The transition elements are put together in the Table because they are similar to each other. Many of the elements common in everyday life are transition elements. You should be able to recognise some of them from the list of symbols given at the beginning of this Topic.

All the transition elements are metals, and so they are similar to look at. You know the physical appearance of a few transition elements from the work you did in Topic A1. You saw that iron and zinc are solids and that mercury is a liquid. All the other transition elements are solids.

This bridge across the Menai Straits is made of iron

As well as looking like each other, the transition elements also behave like each other. For instance, many of them are physically strong (stronger than the other group of metals that you know – the alkali metals). So for instance, we use iron for building bridges, buildings, cars and many other things. Transition metals are also often good conductors of electricity and heat. Electrical wiring is made of copper.

It is fortunate (in view of the uses to which they are put) that the transition elements are different in another way from the alkali metals. *Transition metals are far less reactive than alkali metals.* A bridge made of sodium or potassium would not last very long! Although iron does rust slowly if left exposed to the atmosphere, it can easily be protected. The first iron bridge was made in 1779. It is still standing today, after over 200 years, in the town named after it (Iron Bridge is between Wolverhampton and Shrewsbury in England). Other old iron bridges can be seen in Bristol (the Clifton suspension bridge, built in 1864), between Anglesey and Wales (the Menai Straits bridge, built in 1850), and near Edinburgh (the Forth rail bridge, built in 1890).

When transition elements do react, the substances that are

formed are often coloured. For instance, copper(II) sulphate is usually blue and cobalt(II) chloride is usually pink. Your teacher may show you some substances which contain transition elements. You will see that the same transition element can produce more than one colour. For instance, substances containing copper can be white, light blue, dark blue or green.

One transition element which shows this ability to produce different colours very well is vanadium. Your teacher may give you the following demonstration.

Experiment 12

The following will be needed: powdered zinc; dilute sulphuric acid; ammonium metavanadate; 100 cm³ measuring cylinder.

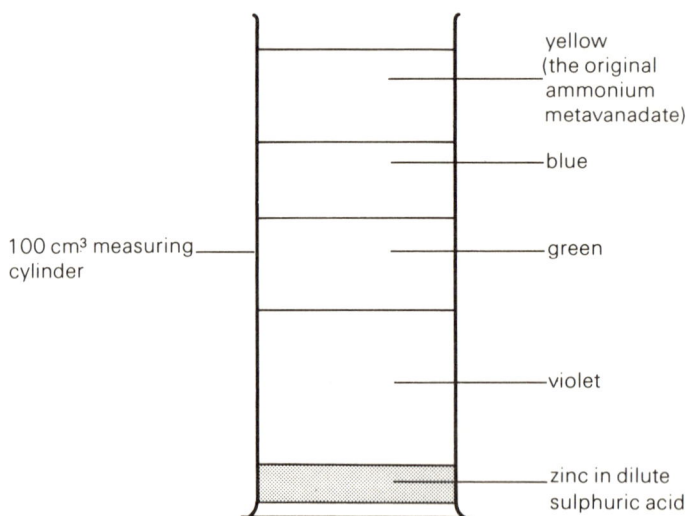

100 cm³ measuring cylinder

yellow (the original ammonium metavanadate)

blue

green

violet

zinc in dilute sulphuric acid

Place a little powdered zinc in the bottom of a 100 cm³ measuring cylinder and cover it with dilute sulphuric acid. To this add about 80 cm³ of a solution of ammonium metavanadate. If this is allowed to stand for a week or more, a series of coloured bands are produced which gradually rise up the measuring cylinder. The zinc reacts with the ammonium metavanadate and produces a number of different substances containing vanadium. Each of these different substances has a different colour.

So now you know about three different types of element – the alkali metals, the halogens and the transition elements. Of these, the alkali metals and the transition elements are metals. The halogens are non-metals. Because they are metals, the alkali metals and the transition elements are similar in many ways.

But as we have seen, they also have many differences. The similarities and differences between alkali metals and transition metals are summarised in the following table.

Similarities	Differences
1 Both types of metal form substances which are bases when they burn in air.	1 Alkali metals are much more reactive than transition metals.
2 They are all solids (except mercury).	2 Alkali metals are much weaker physically than transition metals.
3 They all have shiny surfaces when clean.	3 Alkali metals usually form compounds which are colourless. Transition metals usually form compounds which have a colour.
4 They are all good conductors of both heat and electricity.	

Because of the special nature of transition elements, many of them are used in everyday life. Some uses have already been mentioned, but the table below gives a fuller list.

Transition metal	Main uses	Number of tonnes used each year
Chromium	Decorative, rust-resistant coatings. Production of stainless steel	2.5 million
Iron	Steel production	490 million
Copper	Wiring, piping, coins (mixed with other metals)	7.3 million
Zinc	Protective coat for iron (galvanizing), in batteries, in solder (mixed with other metals)	6.0 million
Tungsten	Production of cutting tools (tungsten carbide)	0.05 million
Gold	Money, jewellery, dentistry, microelectronics	0.001 million

Topic summary and learning

1 Each element has its own symbol, consisting of either one or two letters.

2 Elements which behave in similar ways are found in the same vertical group in the Periodic Table of elements. The only elements which do not belong in vertical groups are the transition metals, which are all similar to each other.

3 Although elements in the same vertical group are similar, there is always a gradual trend in the strength of properties down the group. For instance, sometimes reactivity increases and sometimes it decreases down a group.

Learn the information contained in the Topic Summary.

You should know the symbols of the 29 elements given at the start of the Topic.

You should remember the trends in properties for the group 1 and group 7 elements.

You should know how the transition metals are similar to, and how they are different from, the alkali metals.

Questions

1   Find the element rubidium (symbol Rb) in the Periodic Table. Now answer the following questions about it (in sentences):

a)   Do you expect rubidium to be more reactive or less reactive than potassium?

b)   Could rubidium be cut with a knife? How easily?

c)   How easily would rubidium burn in air? Would it burn easily in chlorine?

d)   When rubidium is added to water, what would you expect to see? What can you say about the solution that would be produced?

2   Find the element fluorine (symbol F) in the Periodic Table. Now answer the following questions about it (in sentences):

a)   Do you expect fluorine to be more reactive or less reactive than chlorine?

b)   Will fluorine react with water? What can you say about the solution produced?

c)   Will fluorine react with iron? How easily do you think it will react?

3   The alkali metals and the halogens will react together. Suppose you had to choose the alkali metal and the halogen that would react most easily. Use your Periodic Table to decide which pair of elements this would be from the following choices:

A.   Cs and Cl
B.   K and I
C.   Li and At
D.   Li and F
E.   Cs and F

4   Which transition metal is worth most per gram?

5   Which transition metal is there most of in

a)   a house
b)   a car
c)   a wedding ring
d)   a ship?

EXPLAINING ELEMENTS

The structure of elements

The elements, which have been the basic theme of this section, have been described as "the building blocks" for all chemical substances. However, a question arises concerning the nature of the building blocks themselves. Of what are they composed? The answer is that the building blocks are extremely small particles called atoms. Each atom in any single element is identical to every other atom in the same element; that is, all atoms of the same element are identical in every respect. What distinguishes one element from another is the nature of the atoms in each of the elements. Different elements are composed of different atoms. For instance, an atom of the element carbon is different from an atom of the element oxygen.

The atoms are far too small to be seen by the human eye, but sophisticated instruments have been used to estimate their size. These show that an atom may be thought of as being spherical, with a diameter of 10^{-10} m! Of course, thinking of an atom as being a very small sphere is simply inventing a model for our understanding of elements, but it is a useful picture for our purposes.

The structure of atoms

The sphere model served science well until Lord Rutherford and others performed a series of experiments in the first half of this century which led to another model in which the atoms themselves had a characteristic structure. It is now believed that atoms are built up from even smaller particles, two of which are contained in a nucleus, and the third in a shell around the nucleus. The particles which are contained inside the nucleus of the atom are known as the proton and the neutron, whilst that in the outside shell is called the electron. The three particles differ in mass and also differ in that two of them are electrically charged. The table below shows the differences between the particles.

Name of particle	Where situated in the atom	Relative mass	Relative charge
Proton	Nucleus	1 unit	+ 1 unit
Neutron	Nucleus	1 unit	Zero
Electron	Shell	approx 1/2000	− 1 unit

Two things emerge from inspecting this table. First of all, most of the mass of an atom will be due to the protons and neutrons it contains (as the electron contributes very little) and will be concentrated right at the centre of the atom in the nucleus. Secondly, because atoms must be electrically neutral overall, the number of protons in the nucleus must be balanced by the same number of electrons in the outer shell. What makes the atoms of one element differ from the atoms of another element is the number of protons (and electrons) which each contains. The number of protons (= the number of electrons) in an atom of an element is called the atomic number of the element. Each element has a different atomic number. The particles contained in a selection of atoms of different elements are given in the table below.

Atom of element	Number of protons	Number of neutrons	Number of electrons	Atomic number of element
Hydrogen	1	0	1	1
Carbon	6	6	6	6
Sodium	11	12	11	11
Aluminium	13	14	13	13
Uranium	92	146	92	92

Atomic number = number of protons = number of electrons

Arrangement of electrons in atoms

It has also been found that the electrons within the atom can themselves be thought of as having a structure. Instead of being arranged randomly around the nucleus of the atom, the electrons are arranged in several shells, rather like the layers around an onion core. A Danish scientist, Neils Bohr, first suggested such a model for the arrangement of the electrons.

In Bohr's model, the electrons are arranged in shells around the nucleus, and the electron structure is built up from the centre shell outwards. So, for the simplest atom, that of hydrogen, the atom could be pictured as shown below.

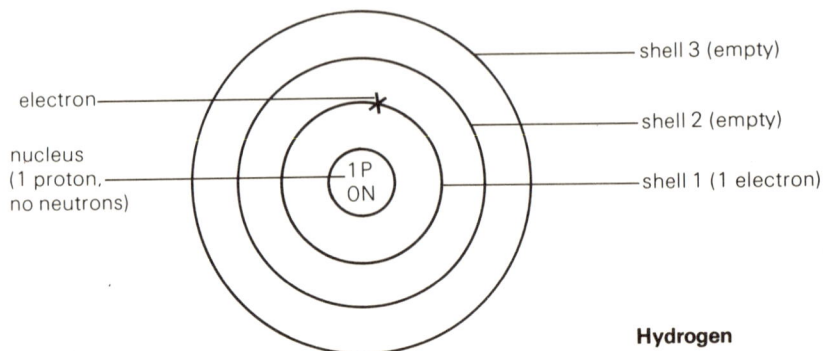

electron

nucleus
(1 proton,
no neutrons)

1 P
ON

shell 3 (empty)

shell 2 (empty)

shell 1 (1 electron)

Hydrogen

The second element (helium) which contains two electrons in each atom, would then be pictured as:

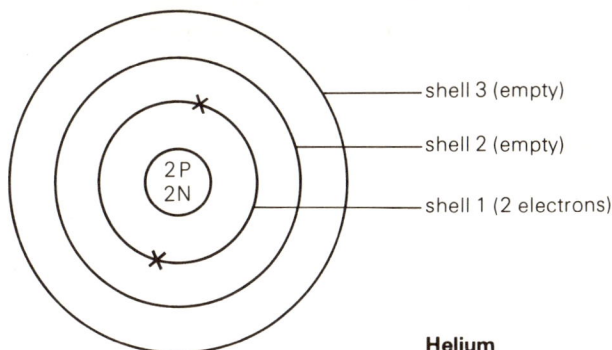

Helium

When it contains two electrons, the first shell is said to be full. Addition of a further electron must take place in the second shell, as shown for lithium:

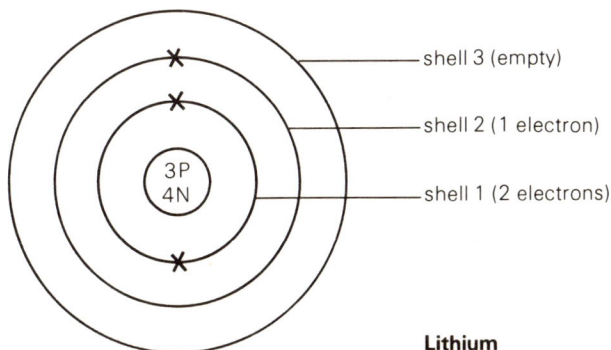

Lithium

The second shell can, in fact, take up to *eight* electrons before it becomes full. A shorthand way of depicting the arrangement of electrons in the shells of twelve atoms is given in the table.

Element (atomic number)	Electrons in each shell			Shorthand way of writing electron arrangement
	1st shell	2nd shell	3rd shell	
Hydrogen (1)	1			1.
Helium (2)	2			2.
Lithium (3)	2	1		2.1
Beryllium (4)	2	2		2.2
Boron (5)	2	3		2.3
Carbon (6)	2	4		2.4
Nitrogen (7)	2	5		2.5
Oxygen (8)	2	6		2.6
Fluorine (9)	2	7		2.7
Neon (10)	2	8		2.8
Sodium (11)	2	8	1	2.8.1
Magnesium (12)	2	8	2	2.8.2

Atomic structure and the Periodic Table

It has already been noted that elements which occur in the same group of the Periodic Table show similar chemical properties. By writing the Periodic Table as in the table below to show the electron arrangement of the elements, it can be seen that elements in the same group of the Periodic Table have the same number of electrons in their outermost shell.

Group 1	Group 2	Group 3	Group 4	Group 5	Group 6	Group 7	Group 8
Li	Be	B	C	N	O	F	Ne
2.1	2.2	2.3	2.4	2.5	2.6	2.7	2.8
Na	Mg	Al	Si	P	S	Cl	Ar
2.8.1	2.8.2	2.8.3	2.8.4	2.8.5	2.8.6	2.8.7	2.8.8
K	Ca	Ga	Ge	As	Se	Br	Kr
2.8.8.1	2.8.8.2	2.8.18.3	2.8.18.4	2.8.18.5	2.8.18.6	2.8.18.7	2.8.18.8

The fact that elements in the same group of the Periodic Table have similar chemical properties, could be supposed to be due to the similar electron arrangement which they have. In particular, it is thought that it is the outermost electrons which are mainly responsible for the chemical properties of an element. It is interesting to note that in group 8, where the so-called rare gases are placed, the outermost electron shell is always completely filled; its electron arrangement is associated with very low chemical reactivity, which is consistent with our knowledge that group 8 elements undergo very few chemical reactions.

Topic 1
ELEMENTS REACTING IN DIFFERENT WAYS

B1.1 Types of chemical reaction

You should already have come across the difference between physical and chemical changes. When a chemical change (chemical reaction) takes place, the following are true:
1   The change usually cannot be reversed easily
2   Heat is usually taken in or given out during the change
3   New chemical substances are formed.

You have already seen many examples of chemical reactions, such as burning elements to make oxides, the reaction of halogens with iron, and the reaction of alkali metals with water. In Section A we used the chemical behaviour of elements to show that they belong to different groups. Different elements (except those in the same groups in the Periodic Table) tend to undergo different sorts of chemical reactions.

When an element reacts it forms a compound. Most elements take part in chemical reactions very easily. The only elements that do not are those at the far right of the Periodic Table (group 8). These elements are so unreactive that for a long time it was thought that it was impossible to make compounds containing any of them. This group of elements are often called the *"inert gases"* ("inert" means unreactive).

To decide whether or not an element has reacted with another element (or with a compound) we can look to see whether statements 1–3 above are true. Often, the most obvious sign that a chemical reaction is taking place is that large amounts of heat are given out or taken in. One of the ways we decided that chlorine was reacting with iron in Experiment 10 was the glow produced in the iron as it reacted. This glow was caused by the heat given out by the reaction. Another example of a reaction between elements that gives out a lot of heat is the reaction between iron and sulphur. This may now be demonstrated by the teacher.

Experiment 13

The following will be needed: weighed sample of iron filings (14 g) and sulphur powder (8 g); a mortar and pestle; hard-glass test-tube.

Mix together 14 g of iron filings and 8 g of powdered sulphur by grinding them in a mortar and pestle. The mortar must be perfectly dry. At this point no reaction will take place. Iron filings can be seen

14 g iron filings
8 g sulphur

grind

mixture
of iron
and sulphur

mixture

HEAT

**examine
contents**

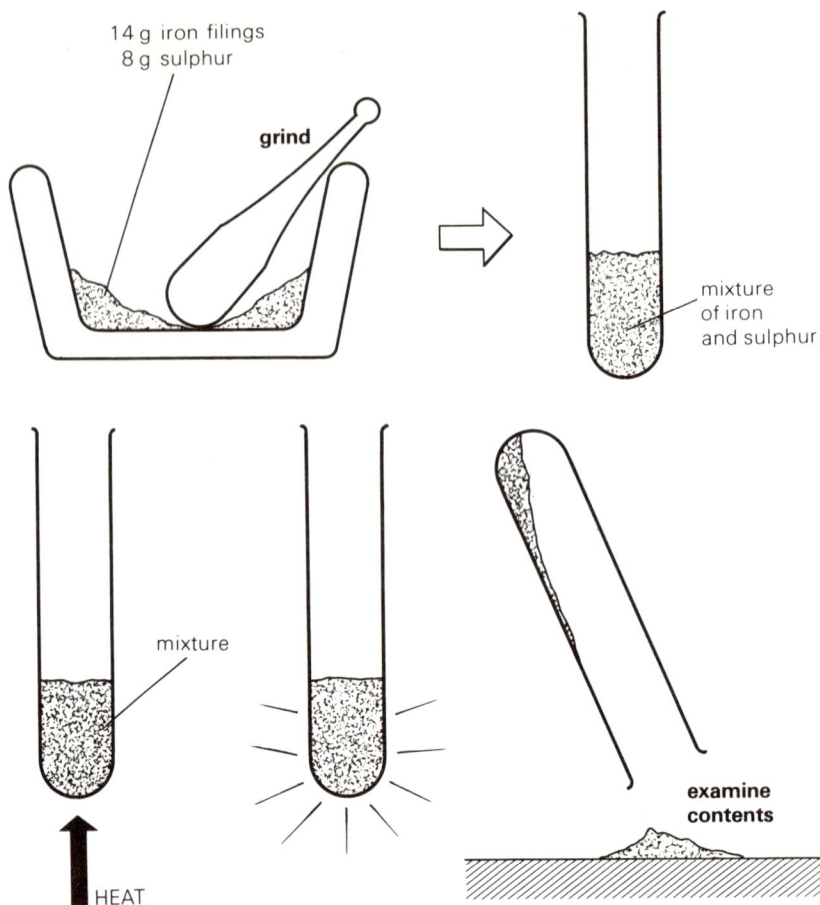

amongst the powdered sulphur, so we have a mixture of the two
elements. Put the mixture into a hard-glass test-tube, and heat the
bottom of the tube in a Bunsen flame. When the mixture begins to
glow, remove the tube from the flame.

After the test-tube has cooled down its contents should be
examined.

During this demonstration you should make observations
which will allow you to note down answers to the following.
1   Does the glow in the tube continue after the heat has been
taken away? Why do you think this happens?
2   Remember what the contents of the tube looked like at the
start of the experiment. Look at the contents again after the tube
has cooled down. Has there been any change during the
experiment?
3   At the start of the experiment a mixture of the elements iron
and sulphur was placed in the tube. What do you think is the
nature of the material removed from the tube at the end of the
experiment? How would you prove this?
4   Make a list of those facts which indicate that a chemical
reaction has take place.

This chemical reaction *gave out* heat. Many chemical reactions do the same. Such reactions are called *exothermic* reactions.

An exothermic chemical reaction is one that gives out heat.

As well as giving out heat, chemical reactions can also take in heat from the surroundings. Reactions of this type are called *endothermic* reactions.

An endothermic chemical reaction is one that takes in heat from the surroundings.

B1.2 Which elements react with which elements?

As already mentioned, the only elements which do not react easily are the inert gases. All other elements react fairly easily forming compounds. This is one of the reasons that so few

The reaction between aluminium and iron oxide is very exothermic: here the heat produced is being used to weld rails together

elements are found as elements in nature. The inert gases are found in nature *only* as elements.

But will any element react with any other element? We could set about answering this question by looking at the different types of element which we know about. We could ask ourselves, for instance, whether metals will react with each other, and whether non-metals react with each other.

Make a list of all the chemical reactions you know of between elements which fall under the following headings:
a)   examples of a metal reacting with another metal
b)   examples of a non-metal reacting with another non-metal.

You should have found it easy to remember examples of metals reacting with non-metals (such as magnesium reacting with oxygen, or iron reacting with chlorine). You also know of cases where non-metals react with other non-metals (such as carbon reacting with oxygen). However, metals do not react with other metals.

Reactions between elements involve a metal reacting with a non-metal, or two non-metals reacting together. Metals do not react with other metals.

B1.3   Representing reactions

Magnesium burns in the oxygen of the air to form magnesium oxide. This can be expressed in the following way:

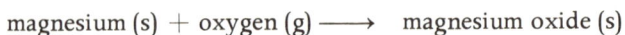

magnesium (s) + oxygen (g) ⟶ magnesium oxide (s)

where (s) is a shorthand way of indicating that the substance is a solid. Similarly, (g) is a shorthand way of indicating that the substance is a gas. When we have a pure substance which is liquid, we use the symbol (l), and when we have a solution of a substance in water, we indicate this by the symbol (aq.). For instance, we can represent the physical change taking place when sodium chloride dissolves in water as follows:

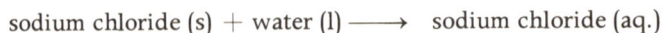

sodium chloride (s) + water (l) ⟶ sodium chloride (aq.)

Since we write out the names of chemicals in words, the above are examples of what we call a *word equation*. The arrow in the word equation tells us the direction in which the change takes place.

For instance, the word equation for the chemical reaction between carbon and oxygen in which the gas carbon dioxide is formed is:

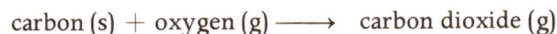

carbon (s) + oxygen (g) ⟶ carbon dioxide (g)

When carbon dioxide reacts with water, the substance which is formed is called carbonic acid. So the word equation for this reaction is:

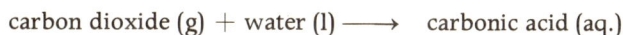

carbon dioxide (g) + water (l) ⟶   carbonic acid (aq.)

Topic summary and learning

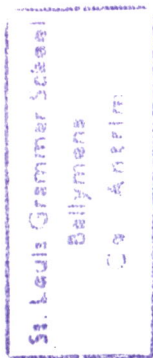

1   When chemical reactions take place, they often give out or take in large amounts of heat.
2   Reactions which give out heat are called exothermic reactions. Reactions which take in heat are called endothermic reactions.
3   Metals will react with non-metals to form compounds, and non-metals will react with other non-metals to form compounds. But metals will not react with other metals.
4   We can represent reactions in shorthand form using word equations.

Learn the information contained in the Topic Summary.

You should remember the characteristic changes that occur when chemical reactions take place. You should know what to look for in order to decide whether a chemical reaction is occurring.

You should learn how to represent reactions using word equations.

Questions

1   Which of the following pairs of elements will *NOT* react together?
a)   A.   Calcium and chlorine
      B.   Iron and chlorine
      C.   Sulphur and oxygen
      D.   Magnesium and sulphur
      E.   Sodium and magnesium

b)   A.   Sodium and bromine
      B.   Potassium and lithium
      C.   Nitrogen and oxygen
      D.   Helium and argon
      E.   Calcium and sulphur

c)   A.   Hydrogen and helium
      B.   Chlorine and oxygen
      C.   Copper and fluorine
      D.   Aluminium and rubidium
      E.   Zinc and iodine

2   Which of the following is an endothermic reaction?

A.   One which occurs between metals and non-metals.

B.   One which takes in heat.

C.   One which occurs between metals and oxygen.

D.   One which gives out heat.

E.   One which occurs between two non-metals.

3   Write word equations to represent the following reactions:

a)   The burning of sulphur in air (the product is called sulphur dioxide).

b)   The reaction between sodium oxide and water (the product is called sodium hydroxide).

c)   The reaction between sulphur dioxide and water (the product is called sulphurous acid).

d)   The reaction between carbon and lead oxide.

e)   The reaction between silver nitrate solution and magnesium metal.

f)   The reaction between carbon and iron oxide.

g)   The burning of lithium in chlorine (the product is called lithium chloride).

h)   The reaction between sodium and water (the products are hydrogen gas and a solution of sodium hydroxide).

i)   The reaction between chlorine and iron (the product is called iron(III) chloride).

j)   The reaction between iron and sulphur (the product is called iron(II) sulphide).

Topic 2
ELEMENTS REACTING AT DIFFERENT SPEEDS

B2.1 Why is the speed of a reaction important?

We have seen that there is a wide variety of types of chemical reaction which take place. In some cases we saw that it was necessary to give the reaction a "push" in order for it to "go". Most often this meant warming the substances together, or shaking them. In this topic we shall be looking at those ways in which we can make chemical reactions go more quickly.

Hydrogen and oxygen react together rapidly to propel this rocket

You already know that some chemical reactions are very slow and others are very fast. Just think of the difference between a piece of wood rotting away in a field and a piece of paper burning in a fire. Apart from the effect which each produces, the time taken for the chemical reaction to be completed is quite different. Can you think of a chemical reaction which is much slower than the wood decomposing? Can you think of a reaction which is much faster than the paper burning? The fact that different reactions take place at different speeds can be made use of when chemicals are produced in industry. It is often important to save time in such processes, because saving time often means saving money.

However, we would be in a much better position if we were able to *control* the speed of chemical reactions, rather than always having to find different reactions which go more quickly. In many cases there will be only one chemical reaction which will suit our purposes. If this is so, being able to change the speed of the reaction will be extremely important.

So we shall investigate the conditions under which a chemical reaction can be made to go faster, or slower. We already have a clue to start with. We noticed that we often have to use heat to get a reaction going. Perhaps the speed of a reaction can always be changed by changing the temperature. We shall choose an element we have already met to see what happens.

B2.2 The effect of temperature on the speed of a reaction

You will carry out some experiments on a chemical reaction at different temperatures.

The reaction which we shall investigate is that between magnesium and dilute acid. Magnesium dissolves in a dilute acid solution to liberate the gaseous element hydrogen. The acid we shall use is dilute hydrochloric acid. We shall see how long it takes for equal quantities of magnesium metal to dissolve in the dilute acid solution at different temperatures.

This is the word equation for the reaction:

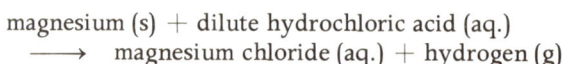

magnesium (s) + dilute hydrochloric acid (aq.)
$\longrightarrow$ magnesium chloride (aq.) + hydrogen (g)

Experiment 14

The following will be needed: 250 cm³ beaker; robust thermometer; stopclock (if available); clean magnesium ribbon (in 1 cm lengths); boiling tube.

a)   Into a boiling tube measure out 20 cm³ dilute hydrochloric acid. Carefully measure the temperature of the acid using a thermometer. Record the temperature of the solution in your laboratory book, in a table like the one opposite.

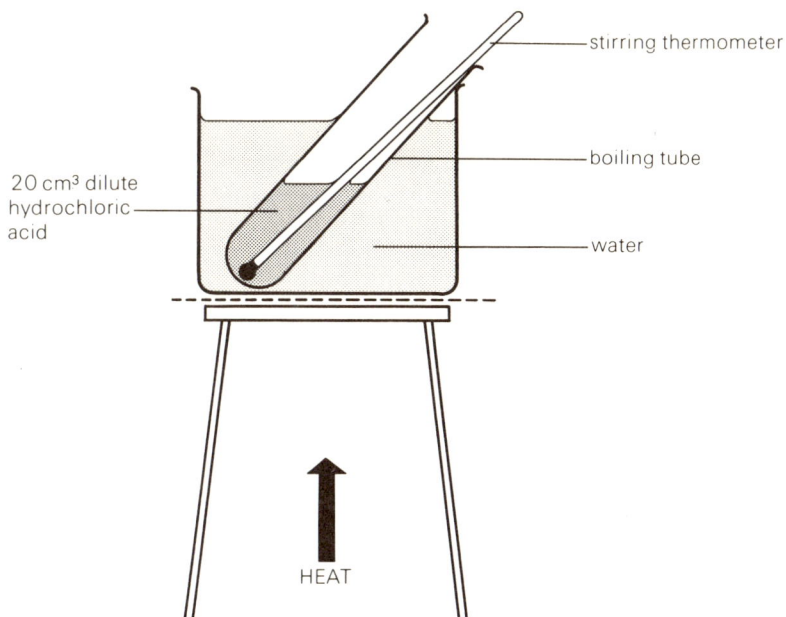

At the same time as you start a stopwatch (or take the time from the seconds hand on a wristwatch) drop into the solution a 1 cm length strip of clean magnesium ribbon. Stir the solution with the thermometer. Notice the bubbles of hydrogen coming from the surface of the metal. When the metal completely disappears, stop the stopwatch (or note the time on your wristwatch). Record in the table the time it has taken for the magnesium ribbon to disappear.

b)   Into a boiling tube measure out 20 cm³ dilute hydrochloric acid. Place the boiling tube in a beaker containing water on a gauze supported by a tripod. Heat the water in the beaker by means of a Bunsen burner. When the temperature of the acid is approximately 25°C remove the Bunsen burner.

Start a stopwatch as you drop a 1 cm length of clean magnesium ribbon into the warm acid. Stir the acid with the thermometer, and note the time when all the magnesium ribbon has disappeared. Record in a table the time taken for the magnesium to disappear. Also record the temperature of the acid.

c)   Repeat the experiment described in (b) for temperatures of approximately 35°C, 45°C and 55°C.

Temperature of acid solution (°C)	Time taken for the 1 cm strip of magnesium ribbon to disappear(s)

In the experiments you always used the same amount of magnesium ribbon and the same quantity of acid. This means that the same "amount of reaction" had to take place each time.

You will notice from the table of results that as the temperature *increases*, the time taken for the magnesium to dissolve *decreases*. Your results can also be put in the form of a graph. Fill in your own results in a similar graph in your laboratory notebook.

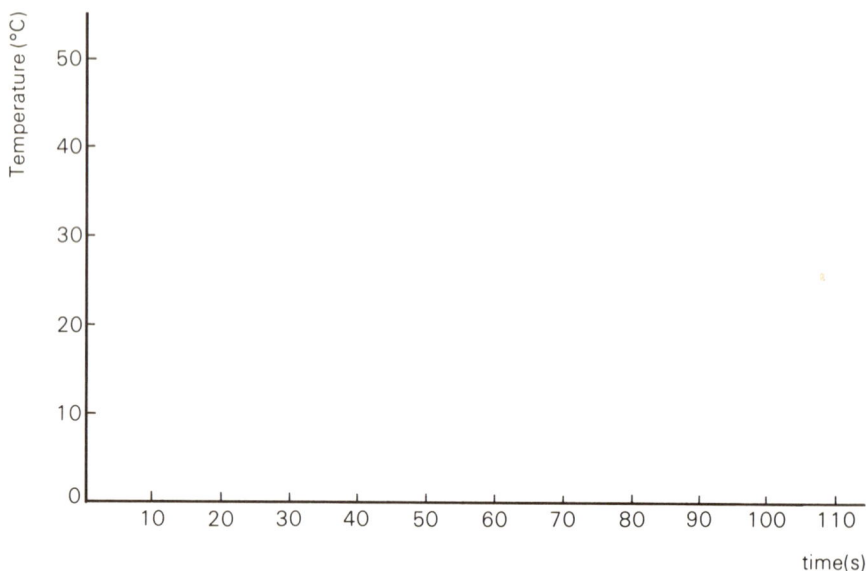

We know that the same amount of reaction took place each time. So the smaller the time taken, the quicker the reaction was going. The reaction went more quickly as the temperature was raised.

An increase in temperature increases the speed of a chemical reaction.

It also follows from the graph that if the temperature of the solution *decreases*, then the time taken for the disappearance of the magnesium will *increase*. This means that *a decrease in temperature will decrease the speed of reaction.*

B2.3 The effect of concentration on the speed of a reaction

You will carry out some experiments on a chemical reaction using solutions with different concentrations.

The reaction which you will investigate is one in which an element is produced by mixing two compounds in solution. The element is sulphur, and it is produced by mixing a solution of sodium thiosulphate with a solution of hydrochloric acid. You can carry out the reaction in a test-tube first of all.

The word equation for the reaction is:

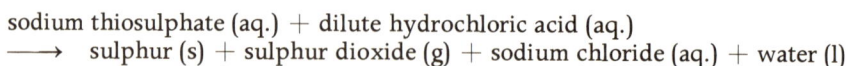

sodium thiosulphate (aq.) + dilute hydrochloric acid (aq.)
⟶   sulphur (s) + sulphur dioxide (g) + sodium chloride (aq.) + water (l)

Experiment 15

The following will be needed: stopclock (if available); 100 cm³ beaker; 150 cm³ sodium thiosulphate solution; dilute hydrochloric acid; two 100 cm³ measuring cylinders.

a)   Pour a solution of sodium thiosulphate into a test-tube until the depth of liquid is about 2 cm. Add about 1 cm³ dilute hydrochloric acid. Observe carefully what happens over a period of time.

You will notice that the sulphur does not appear straight away after mixing the two solutions. It appears gradually, first as a white, milky substance, and then as a yellowish precipitate. It is obvious that the reaction is quite a slow one, because the sulphur takes some time to be produced. Since the solid sulphur is formed in a very light powdery state, it does not settle at the bottom of the test-tube but remains suspended in solution. Because of this the solution becomes more opaque (more difficult to see through) with time. This fact is used to find the speed of the reaction, as explained in the next experiment.

(b)   Into a 100 cm³ beaker put 50 cm³ sodium thiosulphate solution (labelled "0.2M"). Place the beaker and solution on a piece of white paper on which is drawn a thick black cross. This is shown in the diagram.

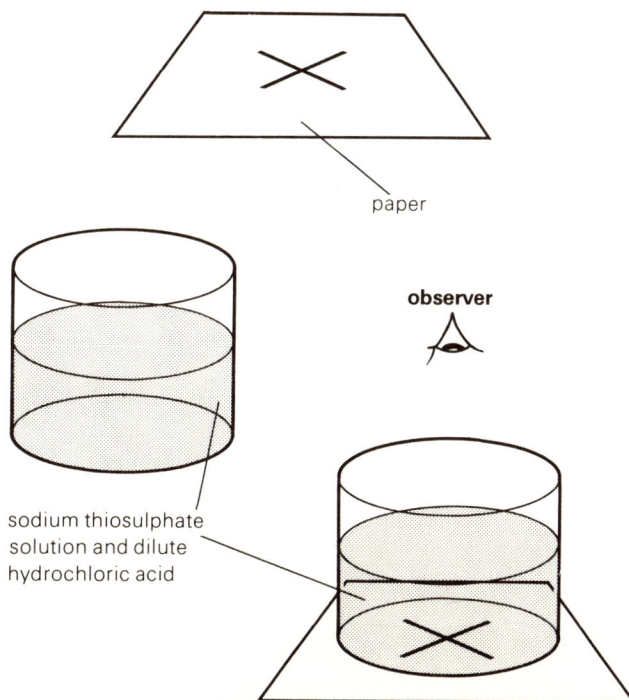

paper

observer

sodium thiosulphate solution and dilute hydrochloric acid

From a measuring cylinder, measure out 5 cm³ dilute hydrochloric acid into a clean beaker. Add the dilute hydrochloric acid to the solution in the beaker and note the time, either by starting a stopclock or by using a wristwatch with a seconds hand. Swirl the contents of the beaker to mix them (be careful not to spill any solution). Observe the cross by looking down through the solution from above. You will notice that as the solution grows more cloudy it becomes more difficult to see the cross. When you can no longer see the cross, stop the stopclock or take the time from the wristwatch. Record the time taken for the cross to disappear.

The experiment can now be carried out with a less concentrated solution of sodium thiosulphate as follows. Put 40 cm³ of the sodium thiosulphate solution labelled "0.2M" into the beaker (washed out thoroughly from the previous experiment). Use a clean measuring cylinder to do this. Add 10 cm³ distilled water to the solution in the beaker and mix the solution by stirring it. The solution in the beaker now has a lower concentration than the solution in the first experiment. Add 5 cm³ dilute hydrochloric acid (previously measured out, as above), start the stopclock and record the time taken for the cross to disappear.

Repeat the experiment for different concentrations of the sodium thiosulphate solution as shown in the table.

Volume of sodium thiosulphate solution (cm³)	Volume of distilled water (cm³)	Concentration of sodium thiosulphate solution decreases	Time taken for cross to disappear (s)
50	0		
40	10		
30	20	↓	
20	30		
10	40		

You will notice that as the concentration of the solution *decreases*, the time for the cross to disappear *increases*. This means that the reaction goes more slowly when the concentration of the solution gets less. It also means that the reaction goes more quickly if the concentration of the solution increases.

An increase in concentration of reactants (the things which react) increases the speed of a chemical reaction.

A decrease in the concentration of reactants decreases the speed of a chemical reaction.

B2.4 The effect of changing the surface area of solid chemicals on the speed of a reaction

You have already seen how the element magnesium dissolves in dilute hydrochloric acid. When it does so, it produces hydrogen gas. Now have a closer look at this reaction.

Experiment 16

a) The following will be needed: 100 cm³ beaker; a 1 cm length of magnesium ribbon.

Into a small beaker put about 20 cm³ dilute hydrochloric acid. Add a strip of clean magnesium ribbon 1 cm long. Observe the release of hydrogen gas carefully. Note particularly where the bubbles of gas come from.

You will carry out some experiments with magnesium metal and dilute acid.

You will notice that the gas is produced at the sides of the magnesium strip. That is to say, the gas is produced *at the surface* of the magnesium. Perhaps if we increased the area of this surface it would allow more bubbles to be produced in the same amount of time. If this happened, we would see an increase in the speed of the reaction. To test this idea you will need to get the same amount of magnesium metal but in two different forms. Use magnesium ribbon and magnesium powder as described below.

b) The following will be needed: access to a top-pan balance; 100 cm³ beaker; clean magnesium ribbon (10 cm); magnesium powder.

1 Weigh a piece of clean magnesium ribbon about 10 cm long. Into a 100 cm³ beaker put 50 cm³ dilute hydrochloric acid. Drop the metal into the acid.
2 Weigh out a small amount of magnesium powder equal in weight to the piece of magnesium ribbon used above (it will be about 0.1 g). Into a 100 cm³ beaker put 50 cm³ dilute hydrochloric acid. Tip the magnesium powder into the acid. Stir with a glass rod to make the magnesium powder mix with the hydrochloric acid.

You will notice that it takes much less time for the powder to dissolve than for the ribbon to disappear. When we grind something into a powder, we give it a much larger total surface area. The surface area of a powder is the surface area of all the specks added together. So the surface area of the magnesium powder which you used was much greater than that of the same weight of magnesium ribbon. The surface area of 0.1 g of magnesium powder is about the same as a desk lid. So the reaction between the magnesium and the acid goes faster when the metal has a larger surface area. This is found to be true for a great many reactions between different metals and acids. It is also true for reactions between solid compounds and solutions of different kinds. (Think how little time a sweet lasts if you break it up by chewing).

An increase in surface area of a solid increases the speed of a chemical reaction involving that solid.

So we have now seen that increasing the temperature, increasing the concentration of the reacting substances (in solution) and increasing their surface area (if solid) will all increase the speed of a chemical reaction. We might expect, therefore, that all reactions done in industry will be carried out with these conditions adjusted so that the maximum speed is obtained. In general this is the case but, often, the use of very high temperatures and concentrations is too expensive to be useful. High temperatures require a lot of money to be spent on heating, for example.

When gases react, the surface area is always very high. (You can't divide substances into particles smaller than the particles in gases.) Instead of talking about the concentration of a gas we talk about its pressure. A high pressure means a high concentration. Increasing the pressure of a gas increases the speed of a chemical reaction in which it is involved. But, as with concentration, it costs a lot of money to use high pressures in industry. So the conditions that are chosen are usually a *balance* between the cost of carrying out the process and the benefit obtained by using high temperatures, concentrations and pressures to increase its speed. There is always a point when it is no longer *worth* increasing the speed any further. We want the chemicals as quickly as possible – but not if it costs too much to make them. Here are some examples of the conditions (temperature, pressure, etc) which are used in particular industrial processes. In each case someone has worked out that these give the best balance between speed and cost.

Name of process	Product	Starting materials	Concentration/pressure/ surface area of starting materials	Temperature
Solvay	Sodium carbonate	Sodium chloride (aq.) Ammonia (aq.) Carbon dioxide (g)	Concentrated Concentrated Under pressure	Room temperature
Haber	Ammonia	Nitrogen (g) Hydrogen (g)	200–500 atmospheres	450°C
Extraction of iron (Blast Furnace)	Iron	Iron(III) oxide (s) Coke (s) Limestone (s) Air (g)	Finely divided and mixed together Under pressure	about 1000°C

Topic summary and learning

1   Increasing the temperature increases the speed of a chemical reaction.

2   Increasing the concentration of the reacting substances in a solution increases the speed of a chemical reaction.

3   Increasing the surface area of solid reacting substances increases the speed of a chemical reaction. Surface area is increased by dividing the solid into smaller pieces.

4   In industry a balance has to be struck between the speed of the reaction and the cost involved in carrying it out. Often, moderately high temperatures (300–400°C) are used because reactions are then often quite quick without being too expensive to carry out.

Learn the information contained in the Topic Summary.

You should remember how you were able to find the effects of changing different conditions on the speed of chemical reactions. In each case you compared the speeds of identical reactions in which you had changed just one condition (such as the temperature). You always needed to carry out more than one reaction so that you could compare their speeds to see what the effect of the change had been.

Questions

1   Which of the following forms of sugar will dissolve most rapidly in hot tea?

a)   ordinary granulated sugar;

b)   lump sugar;

c)   caster (very fine) sugar.

2   Can you explain how you would test your answer to question 1 by doing an experiment? Try to do so in your own words.

3   It takes much longer to cook a meal on a camping stove than on a house stove. Explain why this is so.

4   The temperature of liquid air is about −200°C. Do you think that liquid air will react more or less rapidly with magnesium than ordinary gaseous air will? Explain your answer in your own words.

5   Which of the following sets of conditions will lead to the most rapid reaction between marble granules (a solid) and hydrochloric acid?

	Temperature	State of marble granules	Concentration of acid
A	20°C	Large granules	Dilute
B	20°C	Powdered marble	Dilute
C	80°C	Powdered marble	Dilute
D	80°C	Powdered marble	Concentrated
E	80°C	Large granules	Concentrated

Explain your answer.

6   A chemical factory makes the compound PVC, which is a plastic from which gramophone records are made. The customer (the record company) has placed an order which means that the factory has to work continuously, making at least a tonne of PVC every 40 minutes. But to keep the cost of records down, each tonne must cost less than £65. The manager of the factory knows that he can make PVC quickly and expensively, or slowly and cheaply. He has the following five sets of conditions for the chemical reaction from which he can choose:

Conditions for reaction	Cost of PVC/tonne	Time take to obtain 1 tonne
A	£100	5 minutes
B	£80	15 minutes
C	£60	35 minutes
D	£40	4 hours
E	£20	10 hours

If you were the factory manager, which set of conditions for the reaction would you choose in order to meet the needs of the record company? How much would 100 tonne of PVC made in this way cost?

Topic 3
ELEMENTS REACTING WITH DIFFERENT STRENGTHS

B3.1 Reactivity

We have already used the word reactivity when talking about elements. We used it when we looked at the group 1 elements and again when we studied the group 7 elements. We said that chlorine was more *reactive* than iodine because it *reacts more easily*. It was far easier to get chlorine to react with iron – iodine had to be heated much more strongly.

We also said that potassium was more reactive than lithium because it burned more easily in air. Lithium has to be heated for a longer time before it burns.

So we already have an order of reactivity for some of the elements we know about:

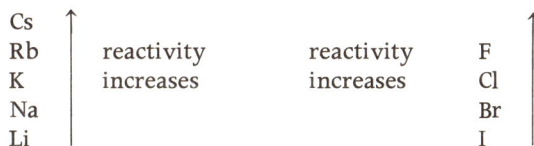

Cs	↑				↑
Rb		reactivity	reactivity	F	
K		increases	increases	Cl	
Na				Br	
Li				I	

But can we find an order of reactivity for the other elements we know about? Suppose we start with the elements sodium, magnesium, iron, copper, gold and silver. Your everyday knowledge of these elements should enable you to write a rough order of reactivity for them. Here is a reminder that will help:

Gold and silver   They both keep their shiny surface for a very long time. We can use silver for making knives and forks, and gold for putting fillings in teeth. Both elements are used to make jewellery.

Sodium and magnesium   They both burn very brightly. We store sodium in oil to keep air and water away from it.

Iron and copper   We use iron for making many things but, if air and water get at its surface, it slowly reacts and forms rust. We also use copper for covering roofs, but it slowly turns green because it reacts with the air.

A rough order of reactivity for these metals is therefore:

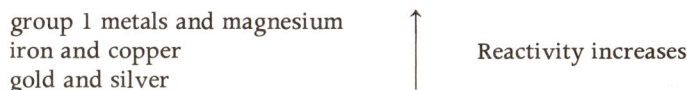

group 1 metals and magnesium	↑	
iron and copper		Reactivity increases
gold and silver		

But we should ask ourselves whether this is accurate enough. Could we make this order more exact? For instance, could we decide which is the more reactive of iron and copper? If we were going to try to do this, we would have to do some

53

experiments – but what would we be looking for? Is there a simple rule that will help us sort out the exact order of reactivity of the elements?

To find this rule we will need to turn our attention away from the metals to the reactivity of a group of non-metals. Think back to the halogens. We know the order of reactivity of chlorine, bromine and iodine, and we know what each of these elements looks like. We can use these facts to help us find the rule that will allow us to sort out an exact order of reactivity for the metals.

B3.2   Reacting halogens with "halides"

You will do some simple experiments in test-tubes.

Halogens are elements. When they react to form compounds, these compounds contain the halogen combined with one (or more) other element. Compounds which contain halogens are called *halides*.

Halides are named after the halogen which they contain. We change the ending *-ine* in the halogen to *-ide* to make the name of the halide.

Name of halogen   *Name of compound (halide) containing it*
Chlor*ine* ───────────→ Chlor*ide*
Brom*ine* ───────────→ Brom*ide*
Iod*ine* ───────────→ Iod*ide*

What would be the name of a compound which contains the halogen fluorine? Where have you heard this name before?

In the experiment which you are about to do, you will be using the following chemicals:
chlorine water – this contains dissolved chlorine (note the colour)
bromine water – this contains dissolved bromine (note the colour)
solutions of potassium chloride, potassium bromide and potassium iodide – halides containing chlorine, bromine and iodine.

We can think of chlorine water and bromine water as being just like the elements chlorine and bromine. They behave as if they were. So we can use them to see how chlorine and bromine would react with the three halides.

Experiment 17

The following will be needed: six test-tubes; a teat pipette; test-tube rack; chlorine water; bromine water; solutions of potassium chloride, potassium bromide and potassium iodide.

Chlorine water and bromine water both give off fumes which are unpleasant. You will therefore find both in the fume-cupboard where they are kept. Be careful not to breathe in any of the fumes from the chlorine water and bromine water you take back to your bench.

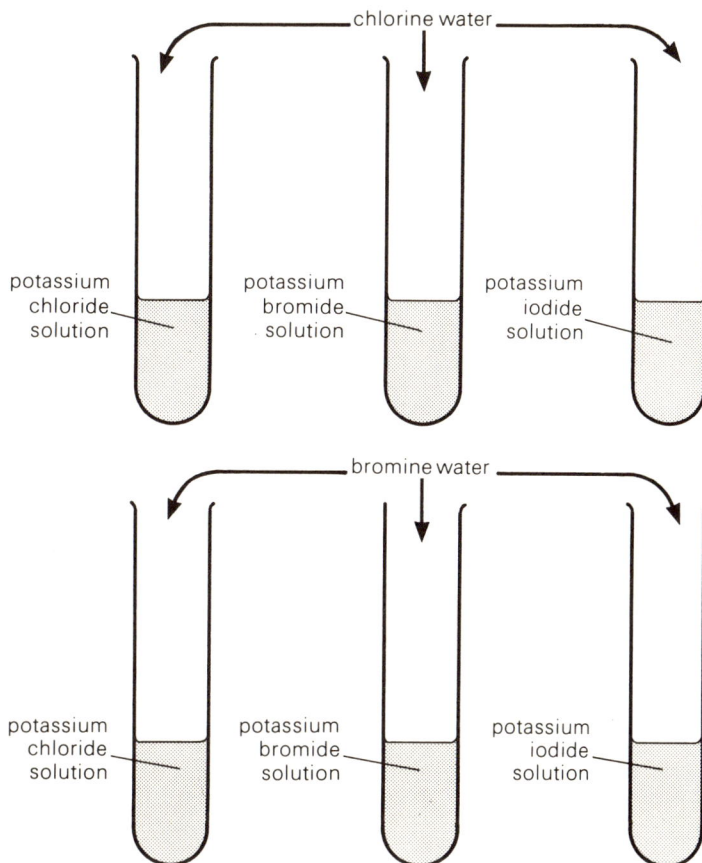

chlorine water

potassium chloride solution   potassium bromide solution   potassium iodide solution

bromine water

potassium chloride solution   potassium bromide solution   potassium iodide solution

You are going to test all three halides with each of the halogens. Make out a table containing all the tests. There will be six in all.

Take three test-tubes and put potassium chloride solution in one, potassium bromide solution in another and potassium iodide solution in the last one. A depth of about 3 cm will be enough in each case. Stand the three test-tubes in a test-tube rack remembering which was which. Using a teat-pipette add a few drops of chlorine water to each test-tube. Shake each test-tube.

If a reaction takes place, there will be a colour change. If there is a colour change, write down what you see.

You know which halogen is in each halide. You know the colours of the halogens. For each case in which there is a reaction, try to say just what is happening. If you cannot, your teacher will help you.

Now repeat the above starting with three fresh test-tubes containing the halides. This time add bromine water with a teat-pipette and shake. Again, make sure you know which test-tube is which.

You should be able to build up a table of results like this:

HALOGEN	Chloride	Bromide	Iodide
Chlorine	No reaction	Reaction Bromine produced	Reaction Iodine produced
Bromine	No reaction	No reaction	Reaction Iodine produced

Now remember that the order of reactivity for these halogens is:

chlorine   ↑   reactivity
bromine | increases
iodine

So when a halogen reacts with a halide, *the less reactive halogen is left as an element at the end.* (Is this true? Look at the table of results above). Chlorine, being more reactive than bromine, reacts with potassium bromide and *bromine is left at the end.* It reacts with potassium iodide *and iodine is left at the end.*

The rule which we can use to help us find an order of reactivity for the metals is:

When elements of different reactivity react in competition with each other the less reactive element is left as an element at the end.

The word equation for one of the reactions you have just carried out is

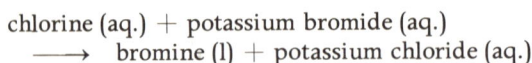

chlorine (aq.) + potassium bromide (aq.)
⟶   bromine (l) + potassium chloride (aq.)

You should write similar word equations for all the other cases where there was a reaction.

B3.3 Finding out an order of reactivity for the metals

You will do a number of simple experiments.

We have found an easy way to compare elements for reactivity. We used non-metals to do this, but what we found works just as well for metals. We took a halogen and tried reacting it with a compound containing another halogen. The less reactive halogen was left as an element at the end.

To compare metals for reactivity we can take a metal and try reacting it with a compound containing a different metal. The less reactive metal will be left as an element at the end. In the experiment which you are about to do, you will be using the following chemicals:

magnesium ribbon lead foil iron filings zinc foil copper foil $\Big\}$ the metals	*Solutions of:* magnesium chloride lead nitrate iron(II) sulphate zinc sulphate copper(II) sulphate silver nitrate $\Big\}$ compounds containing metals

Experiment 18

The following will be needed: microscope slides or white tiles; a glass rod or a teat-pipette; the metals and solutions of metal compounds in the list above.

Do not let the silver nitrate solution touch your skin.

These reactions are carried out on a microscope slide. You will need only a drop of solution and a tiny piece of metal each time.

Take each solution separately, one by one. Place a drop of the solution at each end of three clean microscope slides, so that you have six drops in all. You will have been given tiny pieces of lead, zinc and copper foil, magnesium ribbon and iron filings. Put a piece of each metal (or, in the case of iron, 4–5 filings) separately into each drop, so that you have put each metal into a drop of the solution.

Look to see if a reaction is taking place in each of the drops of solution. If it is, you may see a different metal being produced, or the solution may change colour.

Make a list of what happens for each metal with the solution you use.

Repeat the experiment with a new solution, making sure you use clean apparatus. Do the experiment with as many of the solutions as you can in the time available, and make a list of what happens each time.

You can use your results to build up an order of reactivity for the metals magnesium, lead, zinc and copper. For instance, iron turns out (or "displaces") lead from lead nitrate solution, copper from copper(II) sulphate solution and silver from silver nitrate solution. The less reactive element is left at the end (it is "displaced" from the solution). So iron is more reactive than lead, copper and silver. But iron did not react with zinc sulphate or magnesium chloride solution – so magnesium and zinc are both more reactive than iron. So we can say:

zinc and magnesium
iron
copper, lead and silver

$\uparrow$ increasing reactivity

You can now use your other results to sort out an order of reactivity for all the metals.

The metal reactivity series

Group 1 metals and magnesium
zinc
iron
lead
copper
silver

$\uparrow$ increasing reactivity

The word equation for one of the above reactions is:

iron (s) + lead nitrate (aq.) $\longrightarrow$ lead (s) + iron(II) nitrate (aq.)

You should write similar word equations to represent the other reactions which took place.

B3.4 Where do carbon and hydrogen come in the reactivity series?

You will do experiments to find the position of hydrogen in the reactivity series.

Carbon and hydrogen are both non-metals. Normally we have to consider the reactivity of metals and non-metals separately because these two types of element react in different ways, so we cannot really compare them. However, both carbon and hydrogen will behave like metals under certain circumstances. Because of this, they can both have a position in the reactivity series of metals. They are both common and important elements, so it is useful to know what their positions are.

We can find the position of hydrogen using the same idea that we used in the last section. We can react metals with a simple compound containing the element hydrogen. All acids contain hydrogen, so a simple acid like hydrochloric acid will do.

Remember that the less reactive element is left as an element at the end of the reaction. So, if we put a metal more reactive than hydrogen with hydrochloric acid, what will happen? The element hydrogen will be displaced from the acid, and bubbles of hydrogen gas will be produced.

Experiment 19

The following will be needed: dilute hydrochloric acid; a short piece of magnesium ribbon; zinc foil; a small piece of iron; copper foil.

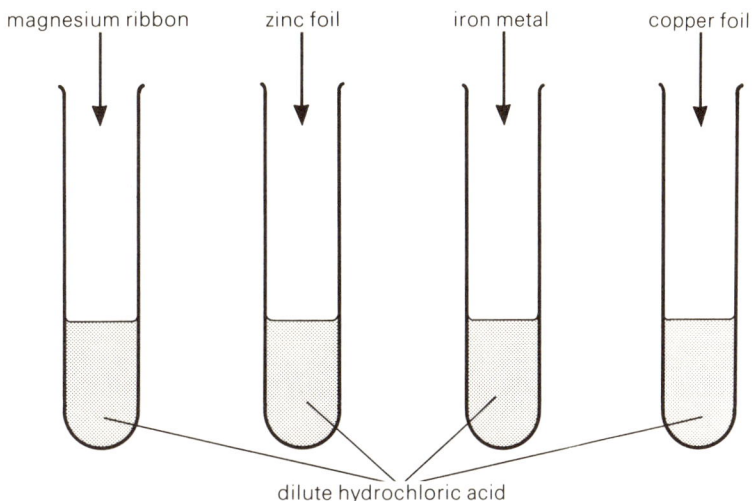

magnesium ribbon   zinc foil   iron metal   copper foil

dilute hydrochloric acid

Put dilute hydrochloric acid in a test-tube to a depth of about 3 cm. Add one of the metals. If you cannot see a reaction taking place, warm the test-tube *very carefully* to speed up the reaction. Do not point the tube towards yourself or anyone else. Keep the tube moving all the time you are warming it.

Try all the metals in turn and in each case decide whether hydrogen is being given off. If you see a gas being produced, put your thumb over the end until you can feel the pressure as the gas builds up. Then quickly test the gas with a lighted splint. Hydrogen burns with a loud pop and sometimes a squeak.

Decide which of the metals displace hydrogen from dilute hydrochloric acid.

You met the word equation for the reaction between magnesium and dilute hydrochloric acid in Topic B2. Look at this again and then try to write similar word equations for the reactions which took place in the experiment you have just carried out.

You should find that magnesium, zinc and iron all displace hydrogen from acids, so hydrogen comes below them in the reactivity series. In fact, *the exact position of hydrogen is just below lead.* In practice, however, it is often difficult to use lead to displace hydrogen from acids.

It is more difficult to find the exact position of carbon from experiments in the laboratory. You know from Topic A2 that

Carbon is used to obtain iron from its oxide in a blast furnace

carbon can be used to obtain metals from their ores. The ores are simple compounds (oxides) containing the metals. The same rule applies as before – it is the less reactive element that is left as an element at the end. *So if a metal can be produced from its oxide by heating the ore with carbon, the metal is less reactive than carbon.*

In theory, we could find the position of carbon by doing simple experiments to see which metals are produced from their oxides by this method. Unfortunately, a Bunsen burner does not give enough heat to start some of the reactions going. We need, instead, to rely on information from experiments that have been carried out at a much higher temperature than we can obtain ourselves in the laboratory. These experiments show that zinc can be obtained from zinc oxide using carbon, but that magnesium cannot be obtained from magnesium oxide. So carbon is more reactive than zinc, but less reactive than magnesium. It therefore comes between magnesium and zinc in the reactivity series.

If carbon is less reactive than magnesium, can you predict what will happen if magnesium is heated with carbon dioxide?

Perhaps if there is time your teacher will show you this reaction. It is easily done by plunging a piece of burning magnesium ribbon into a gas jar full of carbon dioxide.

Knowing the reactivity series lets us predict whether or not a particular reaction will take place. Obviously, this is very useful. It saves us having to try each different reaction one by one to see if it takes place. Like the Periodic Table, the reactivity series saves us time and effort.

Topic summary and learning

1 The reactivity of an element is the strength with which it reacts.

2 When elements of different reactivity react in competition with each other, the less reactive element is left as an element at the end. The more reactive element is left in a compound at the end.

3 The metal reactivity series is:

Group 1 metals
and magnesium
←——————— carbon

zinc
iron   reactivity
lead   increases
←——————— hydrogen
copper
silver

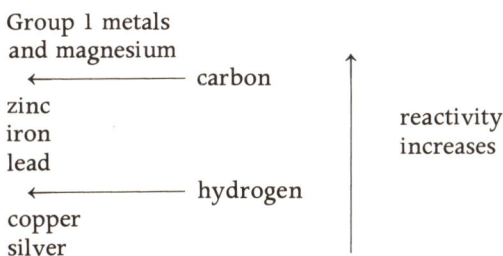

4 A displacement reaction is one in which a less reactive element is displaced from one of its compounds by a reaction with a more reactive element.

5 Compounds containing the halogen elements are called halides (fluorides, chlorides, bromides and iodides).

6 All acids are compounds containing the element hydrogen.

Learn the information contained in the Topic Summary. You should remember the details of the experiments which you have carried out in this Topic. In particular, you should remember how you were able to tell that particular elements (halogens, metals and hydrogen) had been displaced from compounds containing them. You should remember how you tested for the gas hydrogen.

Questions

1 Write down in not more than three sentences what the reactivity series tells us.

2 Make a list of the metals which we can obtain from their ores (oxides) by heating with carbon.

3 Make a list of those metals which will *not* displace hydrogen from dilute hydrochloric acid.

4   Use your knowledge of the reactivity series to say which of the following reactions would take place (there will be more than one):
A.   magnesium + zinc oxide
B.   hydrogen + copper oxide
C.   carbon + sodium oxide
D.   copper + zinc oxide
E.   carbon + silver oxide.

   Use the principle developed during this Topic to say what would be left at the end of each reaction which takes place (*both* products). Then write a full word equation for those reactions which take place.

5   Use your knowledge of the reactivity series to say which of the following reactions would take place (there will be more than one):
A.   bromine + sodium chloride
B.   fluorine + sodium iodide
C.   zinc + dilute nitric acid
D.   iodine + sodium bromide
E.   copper + dilute sulphuric acid.

   Name the elements displaced in those reactions which take place, and write the word equations.

Topic 4
ELEMENTS REACTING IN FIXED AMOUNTS

B4.1 What happens to mass during a chemical reaction?

> *Your teacher will carry out an experiment.*

In this Section you have learned a lot about *how* elements react. But one thing that has not been mentioned is the *amount* of an element that reacts. We do not even know what happens to the total amount of matter during a chemical reaction.

Perhaps some matter is always lost when a chemical reaction takes place. Or maybe the reverse is true – it could be that at the end of a chemical reaction there is always more matter than at the start. So before we look at the amounts of individual elements involved in particular reactions, we must decide exactly what is happening in general to the matter which is involved in chemical reactions.

We can do this in a very simple way. If at the beginning of a chemical reaction we measure the total mass of the things which react together, and then measure the total mass of the things which are produced at the end of the reaction, we can compare the two. By doing this we will find out whether matter has been created or destroyed during the course of the reaction.

The following experiment is best demonstrated by the teacher.

Experiment 20

The following will be needed: access to top-pan balance; 250 cm³ conical flask with tight-fitting bung; small test-tube; sodium chloride solution; silver nitrate solution.

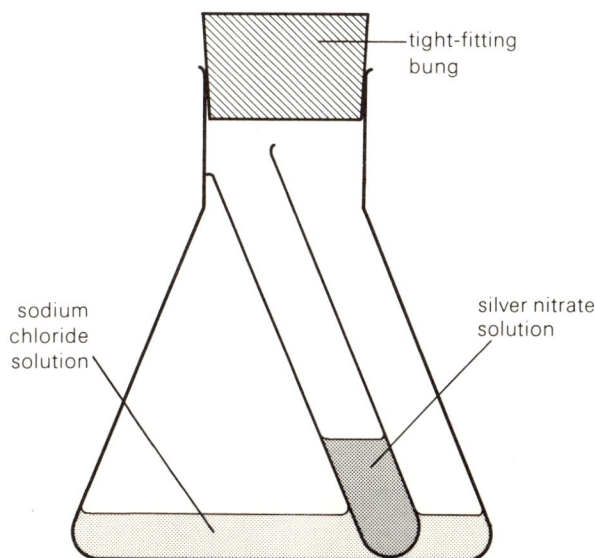

tight-fitting bung

sodium chloride solution

silver nitrate solution

Set up the apparatus as shown. If necessary, tie a piece of cotton around the neck of the test-tube to support it, by catching the cotton between the bung and the side of the flask. Weigh the apparatus on the top-pan balance.

Ensure that the bung is fitted tightly, and then invert the apparatus several times, allowing the solutions to mix completely. You will see that a reaction has occurred because a white precipitate of silver chloride will form. The word equation is:

silver nitrate (aq.) + sodium chloride (aq.) → silver chloride (s) + sodium nitrate (aq.)

Now weigh the apparatus.

During your teacher's demonstration, you should note the following:
1   The weight of the apparatus at the beginning.
2   What happens when the apparatus is inverted. Note down the word-equation.
3   The weight of the apparatus at the end.

You will see that the weight of the apparatus does not change during the course of the reaction. Since the weight does not change, the mass of the products must be the same as the mass of the starting materials. Therefore no matter was created or destroyed when the reaction took place. This is true for *all* chemical reactions.

Matter is neither created nor destroyed during a chemical reaction.

Another way of saying this is that the amount of matter is *conserved* (stays the same) during a chemical reaction. So the rule above is known as the *Law of Conservation of Mass*.

B4.2 Finding out how much of an element reacts

You will do a simple experiment using magnesium and hydrochloric acid.

Now that we know what happens in general to mass during a chemical reaction, we can go on to look more closely at particular examples.

Suppose, for instance, that we were reacting the elements copper and oxygen together to make the compound copper(II) oxide. Suppose, also, that we started with 10 g of copper and measured how much oxygen we needed to react with all the copper, and that we did this experiment several times. Would the amount of oxygen needed be the same each time? Perhaps elements do not react together in the same amounts each time, and so perhaps the amount of oxygen would always be different.

But suppose the amount was always the same – and that it stayed the same from day to day, and even from year to year. And suppose all the elements always react together in fixed amounts. It would mean that there was an important rule which elements always obey.

Can we find such a rule? We can try to discover one by carrying out some experiments.

Instead of trying the reaction between copper and oxygen, we shall use an easier one to handle, the reaction between magnesium and an acid. The question we are going to ask is still the same. Suppose we take a certain amount of acid, and measure how much magnesium we need to react with it. If we have the same amount of acid, do we always need the same amount of magnesium?

You know from Topic A3 that magnesium reacts with dilute hydrochloric acid, and produces hydrogen. As it does this, the magnesium becomes a compound of magnesium which immediately dissolves. So it is easy to see when all the magnesium is used up – no magnesium ribbon remains.

Experiment 21

The following will be needed: measuring cylinder; 250 cm³ beaker; Universal Indicator Paper; magnesium ribbon (10 cm), approximately 0.5M hydrochloric acid (15 cm³).

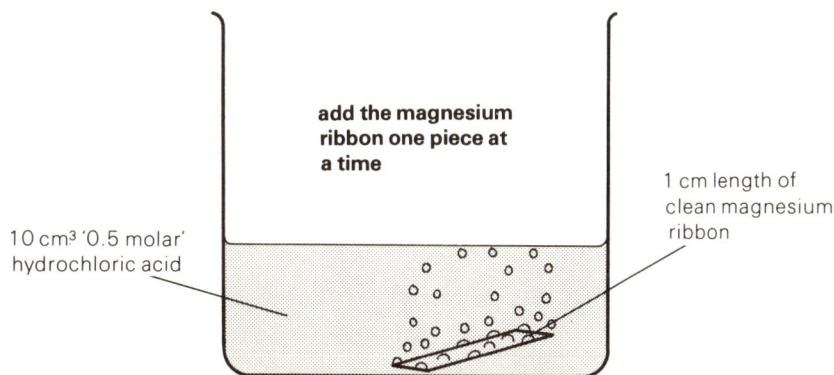

add the magnesium ribbon one piece at a time

1 cm length of clean magnesium ribbon

10 cm³ '0.5 molar' hydrochloric acid

1   You will be provided with some dilute hydrochloric acid marked "0.5 molar hydrochloric acid". Take some of this acid back to your bench in a beaker.

2   Using a measuring cylinder, measure exactly 10 cm³ of the hydrochloric acid into a clean 250 cm³ beaker.

You are going to see how much magnesium you need to react with this amount of acid.

3   Take a piece of magnesium ribbon about 10 cm long and clean its surface with sandpaper.

4   Using a ruler to measure with, cut the ribbon into pieces exactly 1 cm long, and put any ribbon left over to one side. You are going to see how many of these pieces are needed before all the acid is used up.

5   Put a 1 cm long piece of magnesium ribbon into the acid. Swirl the acid around in the bottom of the beaker, but be careful not to let any splash out. Keep swirling, because this helps the magnesium to dissolve more quickly.

6   If the first piece dissolves completely, make a note of this, and then add a second piece. As before, swirl the acid round to speed things up.

7   Each time a piece of magnesium dissolves completely, make a note and then add another one. *Do not try to rush the experiment by adding two pieces together. This will spoil your results.*

8   Eventually, after a few pieces of magnesium have dissolved, the reaction becomes quite slow. How could you speed it up? Think back to Topic B2. The easiest way would be to use heat. Take the beaker and place it on a tripod and gauze. Using a low Bunsen flame, heat it gently. You should try not to let the acid boil. If it does, take the Bunsen away, but make sure you are keeping the temperature fairly high.

9   From time to time, pick up the beaker carefully and swirl the contents.

10   Do not let the beaker boil dry. If this looks like happening, add some distilled water (this does not ruin the experiment, because it does not alter the amount of acid there is).

11   Carry on the experiment, adding a piece of magnesium when necessary, until the reaction stops. You will be left with a piece of magnesium which has not reacted in the solution.

12   You should have a note of the number of pieces of magnesium which were dissolved completely by the acid.

13   Test your final solution with a piece of Universal Indicator Paper. What has happened to the acid?

At the end of the lesson your teacher will collect together everyone's results. These will show that the same amount of magnesium was always needed to react with the acid. At the end of the experiment all the acid had been used up.

If you had started with twice the amount of acid, you would have needed twice the amount of magnesium to use it all up. If it had been three times the amount, you would have needed three times the amount of magnesium.

What we have discovered is an important rule about the amounts of elements which react together. We could put it like this:

If two elements A and B react together, the ratio $\dfrac{\text{mass of A reacting}}{\text{mass of B reacting}}$ *is always the same, no matter how many times the reaction is carried out.*

All elements obey this rule. They always react together in certain fixed amounts (that is, in definite proportions). This rule is often called the *Law of Definite Proportions.*

B4.3 Finding the amounts of elements in compounds

> *You will find the relative amount of hydrogen and oxygen in water.*

If elements always react together in fixed amounts, what does this mean about the things they react together to make? What about compounds?

If the same amounts of the elements concerned always go together to make a certain compound, surely the compound must always contain these elements in the same amounts.

Does a compound always contain its elements in the same fixed amounts? Let's take a simple compound – water. Water is made of two elements, hydrogen and oxygen. But is it always made in the same way?

If we pass an electric current through water, it is split up into hydrogen and oxygen. These are both gases – so we can collect them and see how much there is.

Experiment 22

The following will be needed: electrolysis cell to allow collection of gases; 12 volt supply; supply of "acidified water".

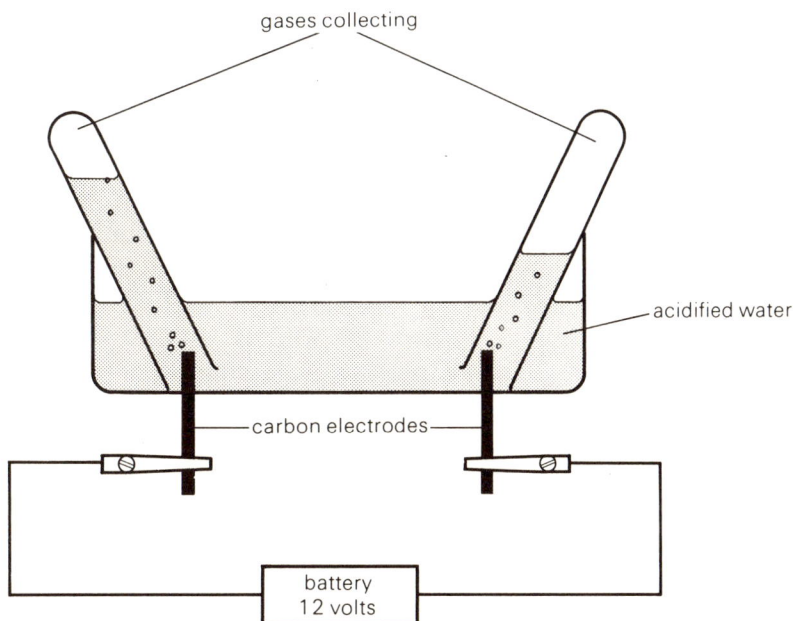

gases collecting

acidified water

carbon electrodes

battery
12 volts

1   You will be provided with an electrolysis cell and some acidified water to pass the current through. (The acid helps the reaction – it does not take part.)
2   Over each electrode place a filled test-tube of water. Your apparatus will probably look like the diagram above.
3   When the first test-tube is half empty, disconnect the battery. Make a rough estimate of the relative amounts of the two gases. Now you have to decide which one is which.

4 Hydrogen explodes with a characteristic pop if you place a lighted splint in it. Oxygen relights a glowing splint. Use these tests to find out which gas is which.

5 Refill the test-tubes and repeat the experiment. Do you still get the same result?

Your teacher will collect together the results from the different groups in the class. These will show that water always contains twice as much hydrogen as oxygen.

Perhaps, if there is time, your teacher will show you an experiment to find the amounts of the elements in another compound. A good example would be ammonia, which is a compound of nitrogen and hydrogen.

All compounds are alike in this way. Each compound contains its elements in the same fixed amounts. No matter how we make a compound, it is always the same. The ratio of the amounts in which its elements are present is always the same.

We could put it like this (strictly speaking, we should talk about the masses of elements):

The same chemical compound always contains the same elements, and the ratio of the masses of the elements is always the same.
(*This is the Law of Constant Composition.*)

Topic summary and learning

1 When a chemical reaction takes place, no matter is created and none is destroyed. The total mass of the starting materials is equal to the total mass of the products. This is known as the Law of Conservation of Mass.

2 When elements react together they do so in definite amounts. The ratio of the masses of two elements which react together is always the same, no matter how many times the reaction is carried out. If the amount of one of the elements is doubled, then the amount of the other element which reacts is also doubled. This is known as the Law of Definite Proportions.

3 Compounds are put together in a regular way. The same compound always contains the same elements. It also contains these elements in a fixed ratio by mass. This is known as the Law of Constant Composition.

4 The Law of Definite Proportions and the Law of Constant Composition are really the same law. They say the same thing in slightly different ways.

Learn the information contained in the Topic Summary. You should learn the laws mentioned in the summary *as they are quoted in the text. You should be able to explain each of these laws in your own words.* You should remember the experimental evidence for these laws which you have seen yourself.

Questions

1 Explain in your own words why the Law of Definite Proportions says the same thing as the Law of Constant Composition.

2 Sulphur reacts with oxygen making the compound sulphur dioxide. 5 g of sulphur react with 5 g of oxygen. How many grams of sulphur would react with
a)   1 g of oxygen
b)   0.1 g of oxygen
c)   12.3 g of oxygen?

3 When carbon dioxide is made from carbon and oxygen, 16 g of oxygen react with 6 g of carbon. What mass of oxygen would react with:
a)   12 g of carbon
b)   3 g of carbon
c)   15 g carbon?

4 There is three times as much hydrogen as nitrogen in the compound ammonia. Some ammonia was split into its elements. 10 cm³ of nitrogen was obtained. How much hydrogen was obtained? (Note: this is a volume, not a mass.)

5 What volume of hydrogen would be needed to react with 10 cm³ of oxygen to make water?

6 When iron reacts with sulphur, 7 g of iron react with 4 g of sulphur. Explain in your own words why 14 g of iron filings were reacted with 8 g of sulphur powder in Section B1, Experiment 13 (p. 37).

States of matter

When many atoms of an element are packed together closely they attract one another, and stick together. They are, nevertheless, constantly moving. So this causes a type of tug-of-war between the tendency of atoms to stick together and the movement which attempts to keep them apart. When the motion of the atoms is very great the attraction of the atoms for one another is overcome almost completely, and the atoms move about with great freedom; this results in a *gas* being formed. A gas is one of the states of matter in which a substance can exist. It arises when the particles which make up the substance are all moving about at relatively high speeds (most of the particles in air travel at about 1400 kilometres per hour). As a result a gas will eventually spread out to fill any container into which it is placed. It cannot therefore be said to have any shape of its own.

When the attractive forces between the particles are more in balance with the tendency of the particles to move apart due to motion, another state of matter, a liquid, will result. The particles in a liquid are therefore not free of each other but move around in clusters. Movement of atoms also occurs within the clusters but the atoms are fairly randomly arranged. This makes a liquid have a definite volume, but since the atoms are still free to move around it does not have a definite shape of its own. Thus a liquid will take up the shape of the vessel which contains it.

When the attractive forces between atoms become very much greater than the motion of the atoms, the other state of matter, a solid, results. Here, motion is limited to vibration of the atoms, which are always in very close contact with one another. The atoms are not arranged randomly but are stacked together in regular ways. This makes the solid have a fixed shape and a fixed volume.

Speed of reaction

Different elements will react together chemically only if their atoms come into contact. There can be no reaction if the atoms of each substance do not collide. This does not mean to say that all collisions will result in reaction – if this was so then all reactions would be extremely fast. There are many factors which determine whether or not a reaction will take place as a result of a collision between atoms but, quite clearly, the more often the atoms of different substances collide the greater likelihood there is of a reaction taking place. So, any effect which can increase the rate at which atoms collide together will be likely to contribute to an increased speed of reaction.

You have already investigated the effect of concentration on the speed of reaction. It was shown that an increase in concentration of reacting substances led to an increase in the speed of reaction. Now, since the concentration of a substance is a measure of the amount of it in a given volume, it follows that if the concentrations of the reacting substances are increased then there is a greater chance of collision between the atoms involved. The effect is similar to that of increasing the number of people on the floor at a dance: the chance of dancers colliding when the floor is crowded is very much greater!

Another way of increasing the chances of collision between reacting atoms would be to speed up the atoms so that they make many more collisions in a given time. This can be done by supplying energy in the form of heat to the substances concerned. When you investigated the effect of temperature on the speed of reaction you found that the reaction was faster at a higher temperature. At the higher temperature, the atoms were moving much more quickly, and colliding more often.

Because the atoms in a solid are fixed in position, it is only the surface atoms which can react at any one time. The atoms within the body of the solid cannot come into contact with atoms of a different substance. It follows that, for a fixed amount of substance in the solid state, there will be a greater chance of the speed of reaction being increased if the surface area can be increased. This is what happened when you carried out the experiment with magnesium and hydrochloric acid. When the magnesium was in ribbon form, the total surface area in contact with the acid was very much less than the surface area of the powdered magnesium. You saw how the increase in surface area led to an increase in the speed of reaction.

Topic 1
DIFFERENT SORTS OF COMPOUND

C1.1 What is a compound?

We mentioned the word "compound" very early in this course – in the very first section on elements. We said then that a compound is a chemical "made up from different elements put together". Read that section again and remind yourself what an *element* is!

In this section we will look more closely at compounds and how they behave – but first we must be very clear about what they are. Our "definition" above is not quite accurate enough, because there is one important thing which we must add to it. A compound *is* made up from different elements put together, *but there must be a chemical reaction between the elements before a compound is formed.*

If there were no chemical reaction between the elements, we would just have a *mixture* of the original elements. Think of some simple compounds that you know, and how they are different in nature from the elements which go to make them up. For instance, think of magnesium oxide:

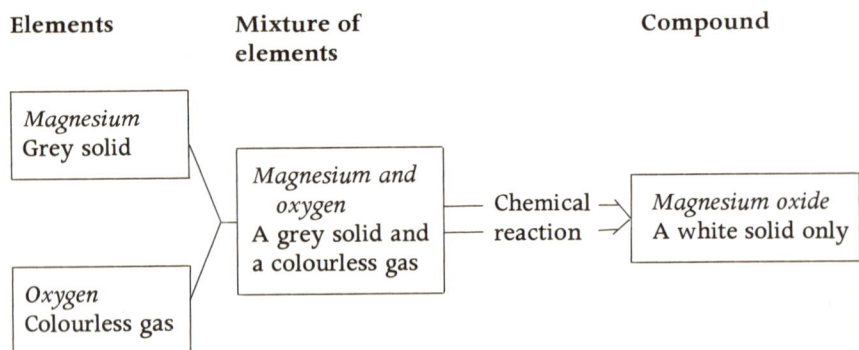

Elements	Mixture of elements		Compound
Magnesium Grey solid	*Magnesium and oxygen* A grey solid and a colourless gas	Chemical → reaction →	*Magnesium oxide* A white solid only
Oxygen Colourless gas			

Obviously, magnesium oxide (a *compound* of the elements magnesium and oxygen) is very different from a *mixture* of the elements magnesium and oxygen. It looks different and if we tested it, it would behave differently. One important difference between a compound and a mixture of elements is that we can easily separate out the elements again if they are in a mixture. It two elements have reacted together chemically to form a compound, it is more difficult to separate out the elements again.

A compound is a substance which is made of two or more elements combined together chemically. It looks different from, and behaves differently from, the elements which make it up. It is a different chemical substance.

Water is a compound of the elements hydrogen and oxygen. How is it different from a mixture of hydrogen and oxygen?

72

We know that there are different types of elements. Are there different types of compounds?

C1.2 Different types of compound

> You will test some compounds to see if they conduct electricity when molten (melted).

We used this test to show that there are different types of element (metals and non-metals). We can use it again here to see if there are different types of compound.

Experiment 23

The following will be needed: power pack or batteries (6 volts); bulb; connecting wires; crocodile clips; carbon rods; lead bromide; ethanol; potassium iodide; paraffin wax.

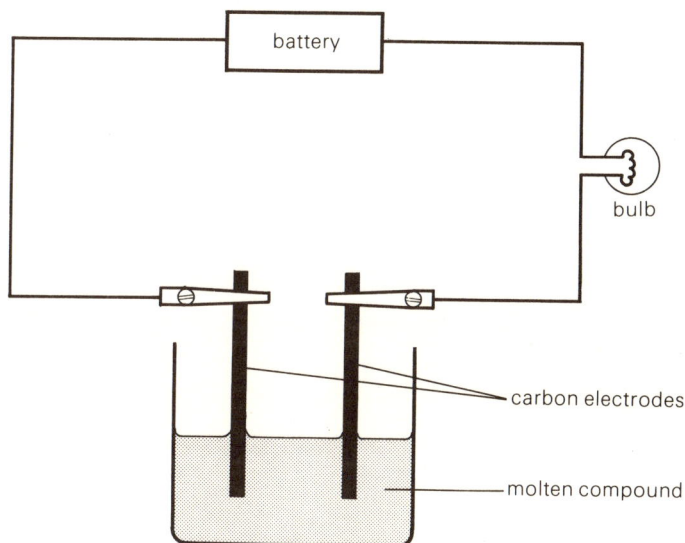

Your teacher will give you four compounds: lead bromide, ethanol, potassium iodide and paraffin wax.

Set up a circuit to test the compounds to see if they conduct electricity.

Ethanol is a liquid and the other three compounds are solids. The three solids have to be melted before they are tested; do this using a Bunsen burner.

Ethanol is very inflammable. Everyone in the room should test ethanol first of all, before anyone has a Bunsen burner alight. When you have finished, do not leave ethanol on your bench. Return the ethanol to a safe place, as instructed by your teacher.

Melt the three compounds in a boiling tube or small beaker. You can then test for conductivity by dipping carbon rods connected to the circuit into the molten compound. Stop heating and move the Bunsen burner to a safe place before you test the compounds to see if they conduct electricity. Remember that the bulb will light up if the rods touch – so be sure that you are testing the compound itself.

Test each compound and write down (next to its name) whether or not it will conduct electricity.

You will find that two of the compounds conduct electricity when in liquid form, and two do not. Why should it be that some compounds conduct electricity while others do not?

Electricity is charge. If something passes electricity it means that a *charge* has been passed. If a compound passes electricity, a charge must have passed between the carbon rods.

But what is it that carries the charge? You should know that all substances are made up of extremely small *particles*. If you have not seen experiments which show this in the past, perhaps your teacher will show you some now. One simple demonstration that shows that matter is made up of particles consists of pouring some ammonium hydroxide solution into a beaker and leaving it on the front bench. The smell of ammonia gas will soon reach those sitting nearby, and will eventually fill the room! The gas ammonia mixes with the air in the room. It can only do this because it is made of tiny particles.

Imagine taking a plastic bag with some red marbles in and adding some blue marbles. If you shake the bag, the marbles will soon be completely mixed together. But if you started with a red block of wood and added a blue block of wood, no amount of shaking would mix them together. Continuous lumps do not mix.

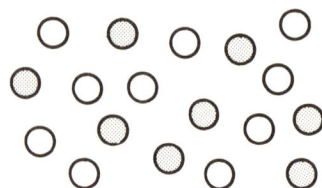

continuous lumps   particles

In the experiment using ammonia, the ammonia could only mix with the air because both were made of particles. So matter is not "continuous", but is made of particles.

When electricity passes through a molten compound, the charge is carried by the particles which make up the compound. In order to be able to do this the particles must be able to move, and they must carry a charge themselves.

One way of making sure that the particles can move is to turn the substance into a liquid. In a solid the particles cannot move, but in a liquid (or a gas) they can.

Solid   Particles cannot move about.
Liquid, gas   Particles can move about.

Some compounds pass electricity when molten. *They are made up of charged particles.* Some compounds do not pass electricity when molten. *The particles which make them up do not have a charge.*

Compounds which are made up of charged particles are called *ionic compounds,* because the charged particles are called *ions.*

Compounds which are made up of particles which have no charge are called *molecular compounds,* because the particles are called *molecules.*

Read through this explanation again.

Ionic compounds conduct electricity when they are in liquid form, but molecular compounds do not. You now know two compounds which are ionic, and two which are molecular. Let us look a bit more closely at the particles which make up the two types of compound – what are ions and molecules?

Ions

There are two types of charge – positive charge, and negative charge (in the same way that there are positive and negative terminals to a battery). If a compound was made up of just one sort of charged particle, it would be either wholly positive or wholly negative. It would be very strongly attracted by a magnet. But if we take an ionic compound like lead chloride and test it with a magnet, nothing happens. *So overall the compound must be neutral.* It must contain positive ions and negative ions, so that their charges cancel out.

*A positive ion is called a cation (Cations are puss*itive*!).*
A negative ion is called an anion.

Ionic compounds are made up of cations *and* anions. But, *overall,* the compound is neutral. When an ionic compound is in liquid form, the ions are free to move about and to conduct electricity. Ionic compounds are those formed when *a metal reacts with a non-metal.* Metal atoms produce the cations and non-metal atoms produce the anions in the ionic compound.

Molecules

Molecules have no charge. A molecular compound contains only one sort of molecule. The smallest particle of an element which can exist on its own is an atom. The smallest particle of a molecular compound which can exist on its own is a molecule. A molecule contains at least one atom of each element which makes up the compound. For instance, the water molecule (H_2O) contains two atoms of hydrogen and one atom of oxygen. *A molecule is itself a particle.* The atoms which make up a molecule are held together very firmly. When a molecular compound is in liquid form, the molecules are free to move about, but they cannot conduct electricity because they do not have a charge.

Molecular compounds are those formed when *two non-metals react together*.

Read through this explanation again.

Topic summary and learning

1   An element is a chemical which cannot be split up into two or more simpler chemicals using a chemical reaction. The particles which make up elements are called atoms.

2   A compound is a substance which is made of two or more elements combined together chemically.

3   There are two types of compound. All compounds are either ionic or molecular.

4   The particles which make up ionic compounds are charged. The positively charged particles are called cations and the negatively charged particles are called anions. Simple ions are atoms of the elements which make up the compound which have been changed so that they have a charge.

5   The particles which make up molecular compounds are *not* charged. The particles are called molecules, and they contain atoms of the different elements which make up the compound held together very tightly. Ions are charged atoms, molecules are many atoms held together.

6   Electricity is charge. For electricity to move, a charge must move. Anything which passes electricity must have something which carries a charge and which can move. Ionic compounds have particles which carry a charge. When an ionic compound is in liquid form, these particles can move. So, if we melt an ionic compound, it will conduct electricity.

7   Molecular compounds do not have anything which carries a charge. So they cannot conduct electricity, even if their particles can move about.

Type of substance	Type of particle	Are the particles charged?
Element	Atoms	No
Ionic compound	Ions (cations and anions)	Yes
Molecular compound	Molecules	No

8   Ionic compounds are formed when an element which is a metal reacts with an element which is a non-metal. Metals produce cations and non-metals produce anions.

9   Molecular compounds are formed when two elements which are both non-metals react together.

Elements   Compounds

```
┌────────────┐
│ Non-metal  │──┐
└────────────┘  │
                │
        Chemical reaction ▷ ┌─────────────┐
                           │  MOLECULAR   │   Particles have no charge
┌────────────┐            └─────────────┘
│ Non-metal  │◁──
└────────────┘
        Chemical reaction ▷ ┌────────┐
                           │  IONIC  │       Particles are charged
┌────────────┐            └────────┘
│   Metal    │──
└────────────┘
```

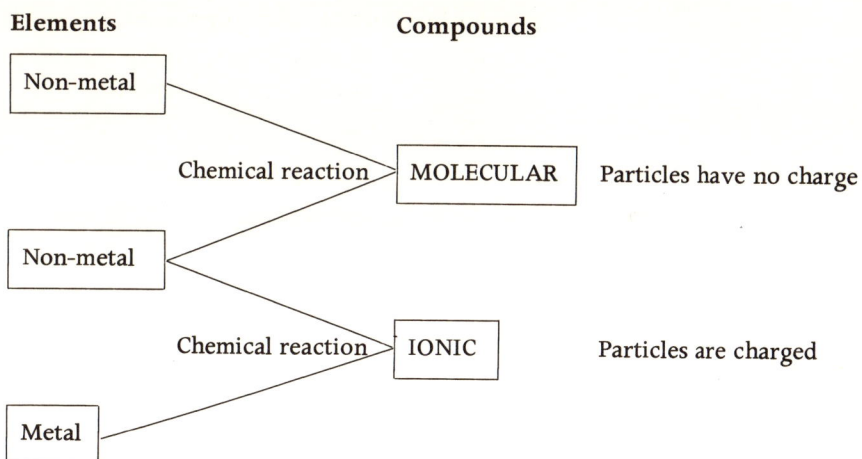

Learn the information contained in the Topic Summary.

Question

Explain in your own words why some compounds do conduct electricity when they are in liquid form while other compounds do not.

Topic 2
DIFFERENCES BETWEEN IONIC AND MOLECULAR COMPOUNDS

C2.1 Do they dissolve in water? Do their solutions conduct electricity?

You will carry out some simple experiments.

We already know one difference between ionic and molecular compounds. When melted, ionic compounds will conduct electricity but molecular compounds will not.

We said that ionic compounds can do this because when they are in liquid form their ions can move. We turned our ionic compounds into liquid form by supplying enough heat to melt them. But there is another way of putting a solid into liquid form. Instead of melting it, we could make a solution of it in some other liquid. But will solutions of ionic compounds conduct electricity in the same way that the molten solids did?

Experiment 24

The following will be needed: power pack or batteries (6 volts); bulb; connecting wires; crocodile clips; carbon rods; sodium chloride; potassium nitrate; ethanol; glucose; potassium iodide; paraffin wax; ethyl acetate; petroleum ether; distilled water.

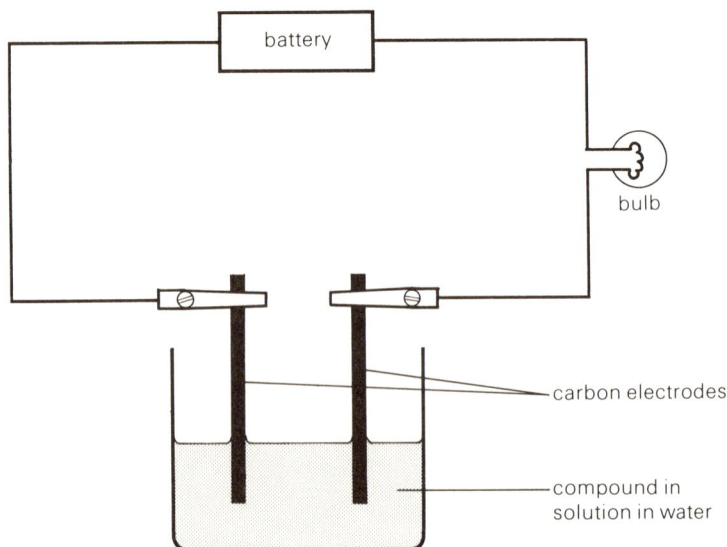

Warning: ethanol, ethyl acetate and petroleum ether are highly inflammable. Keep them well away from any flame.

Your teacher will provide you with a number of compounds from the above list. Some of them are ionic, and some molecular.

Set up a circuit which you can use to test a liquid to see if it conducts electricity (the same circuit that you used in Experiment 23 p. 73).

Take each compound in turn and place a little of it (half a spatula-measure in the case of a solid) in a test-tube. Half-fill the test-tube with *distilled* water and shake it to see if the compound dissolves. For liquids, you should put a depth of about 1 cm in the test-tube, add the distilled water and shake. A liquid is soluble in water if two separate layers are *not* formed when the liquid and water are shaken together.

If the compound you are testing dissolves, pour the solution into a beaker and use your circuit to see if it will conduct electricity.

Draw up a table of soluble and insoluble compounds. Divide the soluble compounds into those whose solution does conduct electricity and those whose solution does not conduct electricity.

Generally speaking, ionic compounds are soluble in water and their solutions conduct electricity. We should expect this to be the case because in the solution ions are still present. Because the particles are free to move in a liquid, the ions of a dissolved ionic compound can move through the solution, conducting electricity.

Some molecular compounds are soluble in water, but most are not. Those few molecular compounds which are soluble in water do not conduct electricity. They have no charged particles.

Look back at your table of results from the last experiment. For each compound, decide whether it is an ionic or a molecular compound. Make a list of those compounds which belong to each type.

C2.2 Differences in melting point and density

Here is some information about five typical molecular compounds and five typical ionic compounds.

Molecular compounds		
Name	*Melting point (°C)*	*Density (g cm^{-3})*
Ammonia	−78	0.77×10^{-3}
Carbon dioxide	−57	1.97×10^{-3}
Ethanol	−117	0.789
Methane	−183	0.72×10^{-3}
Sulphur dioxide	−76	2.93×10^{-3}

Ionic compounds		
Name	*Melting point (°C)*	*Density (g cm^{-3})*
Ammonium chloride	335	1.53
Lead chloride	501	5.90
Potassium nitrate	334	2.10
Sodium chloride	804	2.16
Magnesium oxide	2800	3.65

You will be given a chart showing the melting point and density of some compounds. The names of five of the ten compounds have been put on the chart, but only the position is marked for the other five.

Use the information above to decide which cross stands for which compound, and then write the name of the compound (and whether it is ionic or molecular) next to the cross which shows its position. When the chart is complete it will show the position of five ionic and five molecular compounds.

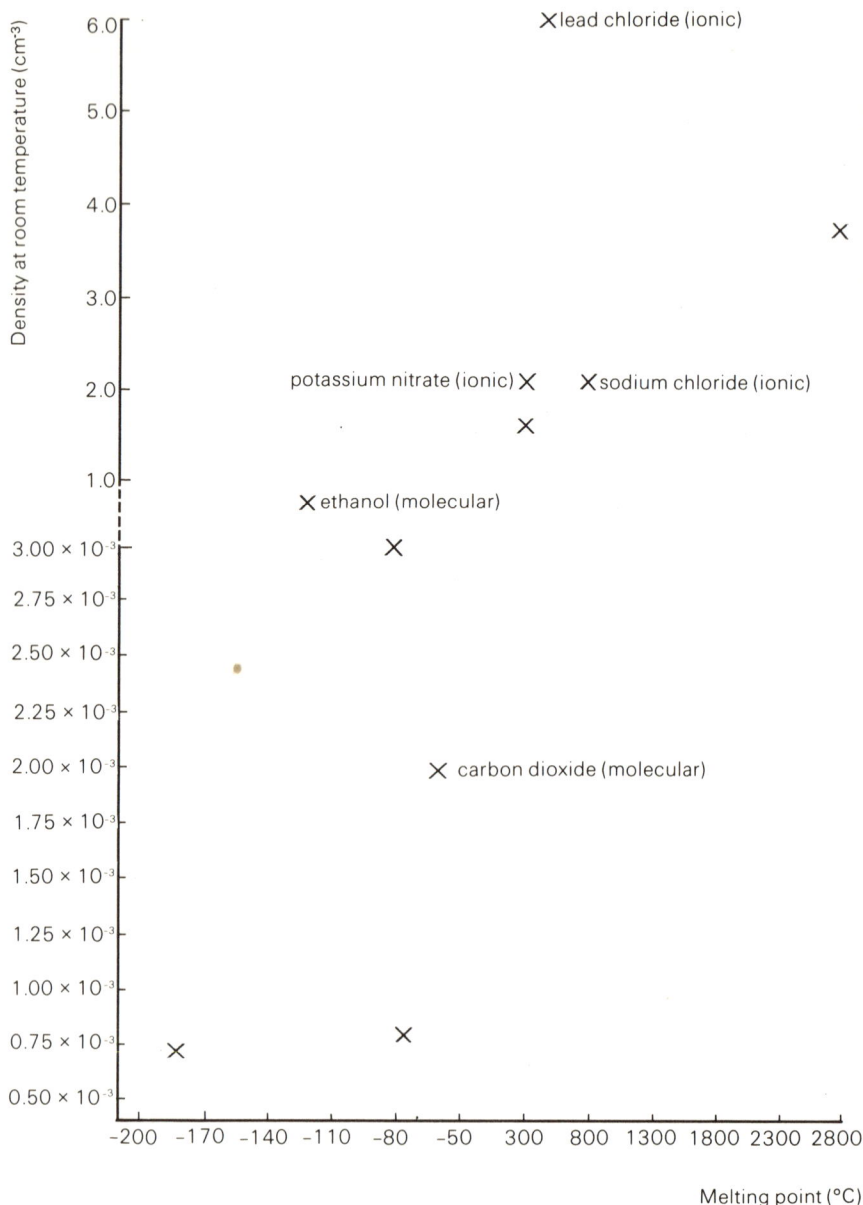

Use the chart you have constructed to answer the following questions:
1   Do ionic compounds tend to have high or low melting points?
2   Do ionic compounds tend to have high or low densities?
3   Do molecular compounds tend to have high or low melting points?
4   Do molecular compounds tend to have high or low densities?
5   Can you summarise this difference between ionic and molecular compounds into one sentence?

C2.3 Differences in speed of reaction

Your teacher will demonstrate that ionic and molecular compounds tend to react at different speeds.

Your teacher will demonstrate a typical reaction between molecular compounds on the front bench. In this reaction the compound benzoic acid is made from the compound benzyl alcohol. The names themselves are not important – what you should notice is how long the reaction takes.

The following experiment must be demonstrated by the teacher.

Experiment 25

a)   The following will be needed: "Quickfit" pear-shaped flask and reflux condenser; anti-bumping granules; test-tube rack; concentrated sodium hydroxide solution; potassium permanganate crystals; benzyl alcohol (phenylmethanol); concentrated hydrochloric acid.

Put 10 cm³ distilled water, 10 cm³ concentrated sodium hydroxide solution and two spatula-measures of potassium permanganate crystals into the pear-shaped flask. Shake thoroughly taking care not to splash any of the mixture. Add some anti-bumping granules.

Clamp the flask in a vertical position and partly submersed in a beaker of water on a tripod. Fix the reflux condenser into the top of the flask and maintain a steady flow of water through it. Add 1 cm³ of benzyl alcohol down the condenser from a dropping pipette.

Heat the water in the beaker until it boils. At intervals of five minutes, withdraw 2–3 cm³ of the reaction mixture from the flask using a dropping pipette, by carefully disconnecting the reflux condenser. To this, add 5 cm³ of concentrated hydrochloric acid in a test-tube. Quickly replace the reflux condenser and continue heating.

If any benzoic acid is present, it will form as a white precipitate as the mixture in the test-tube cools. The product is formed by the reaction in the pear-shaped flask – adding hydrochloric acid simply allows us to see it.

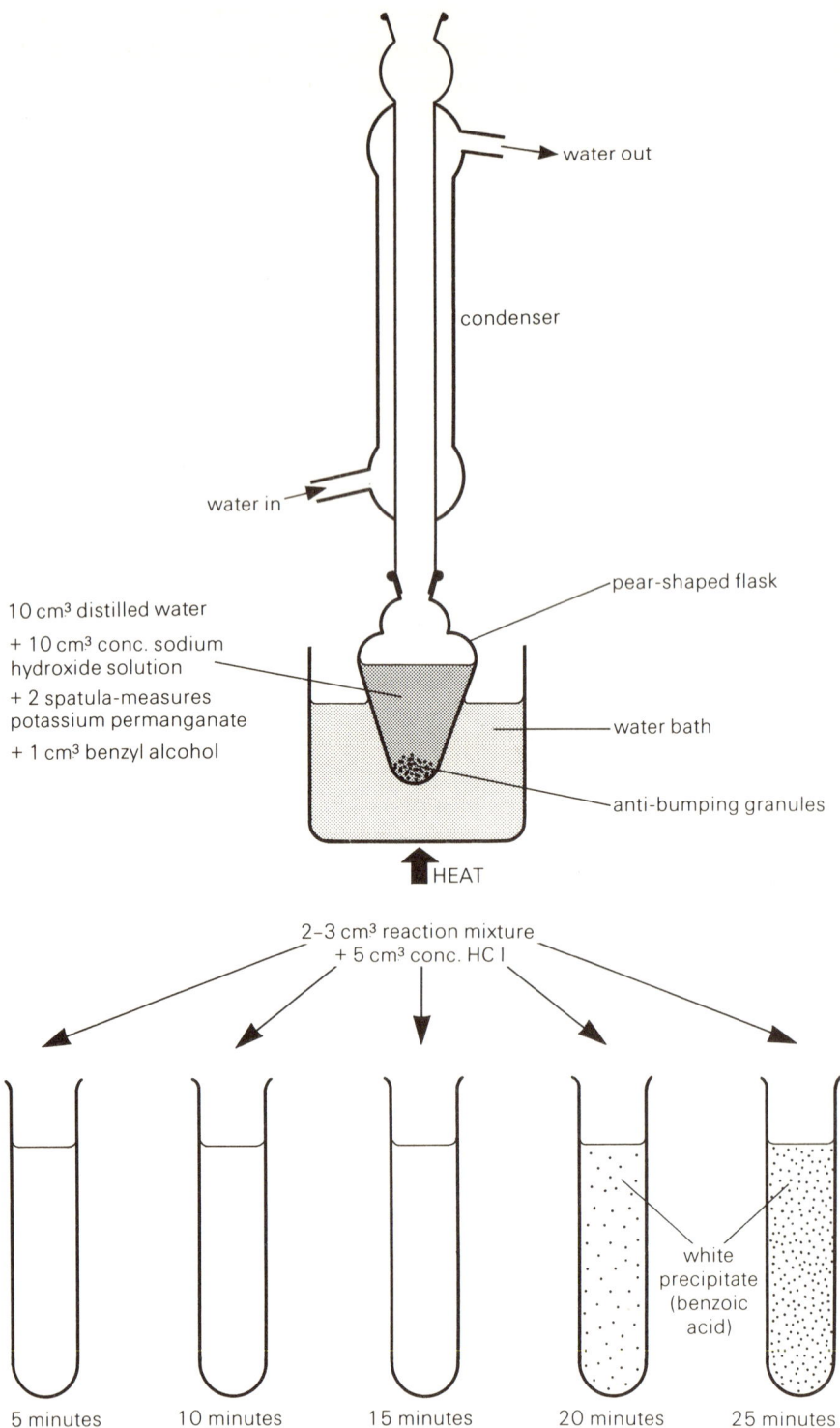

Keep the test-tubes on display on the front bench as shown. It will be about 25 minutes before the reaction produces appreciable quantities of benzoic acid.

During the time taken for this reaction involving a molecular compound, your teacher will show you some typical reactions involving ionic compounds.

b)   The following will be needed: test-tube rack; solutions of sodium chloride, barium nitrate, lead nitrate, silver nitrate, sodium sulphate, potassium iodide and potassium chromate.

While the following reactions are being carried out in test-tubes, pupils should note the speed with which the products (in each case a precipitate) are formed.

1   Sodium chloride solution (3 cm³) + silver nitrate solution
(5 drops).

2   Barium nitrate solution (3 cm³) + sodium sulphate solution
(5 drops).

3   Lead nitrate solution (3 cm³) + potassium iodide solution
(5 drops).

4   Barium nitrate solution (3 cm³) + potassium chromate solution
(5 drops).

Compare the time taken for the reaction of the molecular compound (about half an hour), with that for the ionic compounds. Can we say that molecular and ionic compounds tend to react at different speeds?

Topic summary and learning

1   Ionic compounds conduct electricity when molten and when dissolved in water. They are nearly all very soluble in water. Ionic compounds tend to have high melting points (and boiling points) and densities. Reactions of ionic compounds are usually very rapid.

2   Molecular compounds do not conduct electricity when molten or when in solution. Their solubility in water is usually very low. Molecular compounds usually have low melting points (and boiling points) and densities. The reactions of molecular compounds are usually much slower than those of ionic compounds.

Learn the information contained in the Topic Summary. In Experiment 25(b) you saw four ionic compounds being formed which were insoluble in water. These were: silver chloride (white), barium sulphate (white), lead iodide (yellow) and barium chromate (yellow). You should remember that these ionic compounds are insoluble, and you should remember their colours.

Questions

1   The compound lithium chloride has a high melting point and it is soluble in water. Do you think it is an ionic or a molecular compound? What other properties is it likely to have?

2   The compound carbon monoxide has a very low density and its reactions are usually slow. Do you think it is an ionic or a molecular compound? What other properties is it likely to have?

3   A solution of the compound potassium fluoride in water conducts electricity. What does this tell you about the compound?

4   Suppose you are given two unknown compounds – compound A and compound B. You have to decide for each whether it is ionic or molecular. Describe tests you could do in the laboratory using simple equipment which would allow you to decide. Explain how you would use the results of your experiments to solve the problem.

5   Use your chart showing melting points and densities to decide whether the following compounds are ionic or molecular:

a)   Copper chloride   (m.p.   498°C; density 3.53 g cm^{-3})
b)   Methyl acetate   (m.p.   −99°C; density 0.93 g cm^{-3})
c)   Isopentane   (m.p. −160°C; density 0.62 g cm^{-3})
d)   Calcium phosphate (m.p. 1670°C; density 3.14 g cm^{-3})

6   Write word-equations to represent the reactions which took place in Experiment 25(b). (You know the name of the insoluble product in each case. Work out the name of the other product, which is soluble, then write each word-equation).

EXPLAINING COMPOUNDS

At the end of Section A, a model of the atom was developed in which it was proposed that the atom consisted of a heavy central nucleus, containing protons and neutrons, and that this was surrounded by successive shells containing the electrons. Then it was suggested that the electrons, and in particular the outermost electrons, are responsible for the chemical properties of the element.

You have seen in this section that when atoms react together to form compounds they do so in one of two main ways. They form either an ionic compound or a molecular compound, each type having distinctive physical and chemical properties. Through the model of the atom which has been proposed, it is possible to account for each of these compound types.

It has been suggested that chemical inactivity is associated with a complete outermost electron shell in the atom. For this reason the rare gases do not have many reactions (see Section A). If a complete outermost shell is a particularly stable arrangement then it might be expected that atoms would tend to change their electron arrangement to achieve a complete outer shell. This is, in fact, what happens. When elements react, they do so to obtain a filled outermost shell of electrons.

An atom with an incomplete outer shell can, in principle, achieve a completion by one of two methods. It can gain more electrons to become complete, or it can lose electrons to leave a complete shell. For example the chlorine atom, with an electron arrangement 2.8.7 could, in principle, achieve a complete outer shell either by gaining an electron to become 2.8.8, or by losing seven electrons to become 2.8. Now, in practice, it will be much less energetically demanding for the chlorine atom to gain a single electron than it will for the atom to lose seven electrons. Atoms of elements with a few too few electrons obtain a filled shell by gaining electrons. Think about the relationship between atomic structure and the Periodic Table. Which type of elements have a few too few electrons – metals or non-metals? For chlorine the change

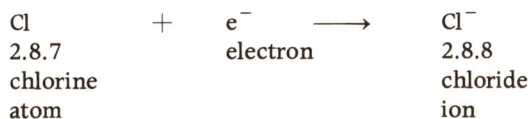

Cl	+	e$^-$	$\longrightarrow$	Cl$^-$
2.8.7		electron		2.8.8
chlorine				chloride
atom				ion

will take place. When the chlorine atom has gained an electron it will have a charge imbalance, since it will still possess 17 protons in the nucleus but will now have 18 electrons in the outer shells. It will therefore become a negative ion, an anion. Other non-metals do likewise, forming negative ions in ionic compounds.

The sodium atom, electronic structure 2.8.1, will prefer to lose one electron to leave a complete shell (2.8) rather than gain seven electrons to achieve completion (2.8.8). Atoms which have a few too *many* electrons obtain filled shells by losing the small number of electrons in their unfilled outermost shell. Which type of element – metals or non-metals – will have a few too many electrons? For sodium, the change which takes place is

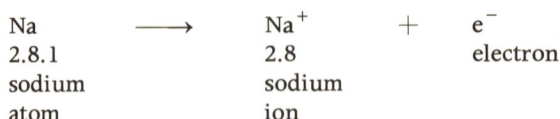

$$Na \longrightarrow Na^+ + e^-$$

Na		Na^+		e^-
2.8.1		2.8		electron
sodium		sodium		
atom		ion		

and the ion will be positively charged (a cation) since it will contain 11 protons in the nucleus but only 10 electrons in the outer shell. Other metals behave in a similar way, forming positive ions in ionic compounds.

It can now be seen that the elements sodium and chlorine would be likely to react with each other, since the sodium atom will become more stable by the *loss* of an electron, and the chlorine atom will become more stable by the *gain* of an electron. This does, in fact, happen.

Na	+	Cl	$\longrightarrow$	Na^+	Cl^-
2.8.1		2.8.7		2.8	2.8.8
				sodium	chloride

The resulting compound, sodium chloride, is composed of ions, and is therefore known as an ionic compound.

Magnesium and oxygen react together by a transfer of electrons:

Mg	+	O	$\longrightarrow$	Mg^{2+}	O^{2-}
2.8.2		2.6		2.8	2.8
				magnesium	oxide

giving rise to an ionic compound. When an element with a few too few electrons (a non-metal) reacts with an element with a few too many electrons (a metal), an ionic compound is formed.

An alternative method of obtaining a filled outer shell for elements with a few too few electrons can be by sharing electrons between shells of different atoms. Take hydrogen and chlorine as an example. Hydrogen (1) needs to gain one electron to achieve a complete first shell and chlorine (2.8.7) also needs to gain an electron to achieve a complete outer shell. The sharing process can be illustrated as follows:

hydrogen atom   chlorine atom   hydrogen chloride molecule

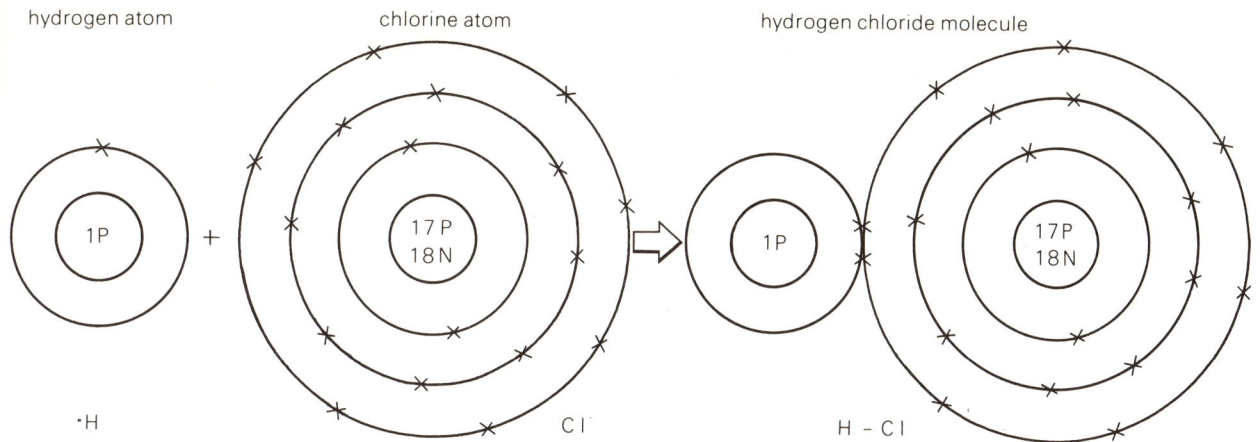

·H   Cl·   H – Cl

Because the hydrogen atom now has a share in two electrons in its outermost shell and the chlorine atom has a share in eight electrons in its outermost shell, each atom within the resultant molecule is more stable than in isolation. Such electron sharing always results in a molecule being formed. The bond between the atoms is said to be a covalent bond.

Many single atoms combine with atoms of the same element to form molecules by an electron-sharing process. For example, the halogen fluorine:

fluorine atom   fluorine atom   fluorine molecule

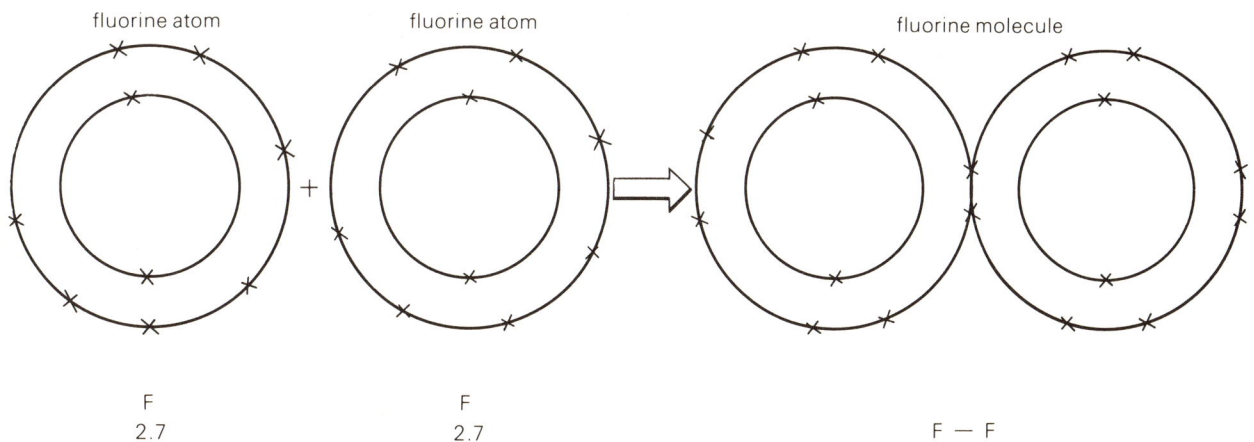

F   F   F — F
2.7   2.7

Many other non-metal atoms do this. This is why many metals are molecular elements. How many non-metal elements can you think of? Try to work out in each case how both atoms obtain a filled outer shell by electron sharing. As we have seen, metals have few electrons in their outermost shell, and will therefore react in order to lose them. It follows that metals always react to form ionic compounds. Non-metal atoms can obtain filled outer shells by electron gain or by sharing, and therefore give rise to both ionic and molecular compounds. So we have seen that:

a)   an ionic compound will have been formed by electron transfer between a metal and a non-metal

and

b)   a molecular compound will have been formed by electron sharing between non-metal atoms.

Section D
Molecules

Topic 1
THE IMPORTANCE OF THE MASS OF A MOLECULE

D1.1 Which everyday substances are molecular?

In the last Section we learned that molecular compounds are formed when two non-metals react together, and that they usually have low melting and boiling points. Molecular substances are likely to be gases or liquids at room temperatures, because there will be enough heat to melt or boil them.

So we can begin looking for molecular substances by considering the everyday liquids and gases which we know. One of the most common chemical compounds in the world is water. We know that at room temperature water is a liquid. This is because it is a molecular substance.

The gases of the air are also molecular substances. Nitrogen, oxygen and carbon dioxide are all molecular. But they are not all compounds! As well as some compounds being molecular, some *elements* are also molecular. A molecular element is one in which two or more atoms of the *same* non-metallic element join together to make molecules. A nitrogen molecule (we write its formula as N_2) is two nitrogen atoms joined together. *It is still an element because there is only one type of atom present.* An oxygen molecule (O_2) is two atoms of oxygen joined together. Again, there is only one type of atom, so oxygen molecules are still the element oxygen.

You know several other *diatomic* (*two*-atom) elements. For instance, the following are all diatomic *molecular elements*:

hydrogen (H_2)
fluorine (F_2)
chlorine (Cl_2)
bromine (Br_2)
iodine (I_2)

Chemical reaction

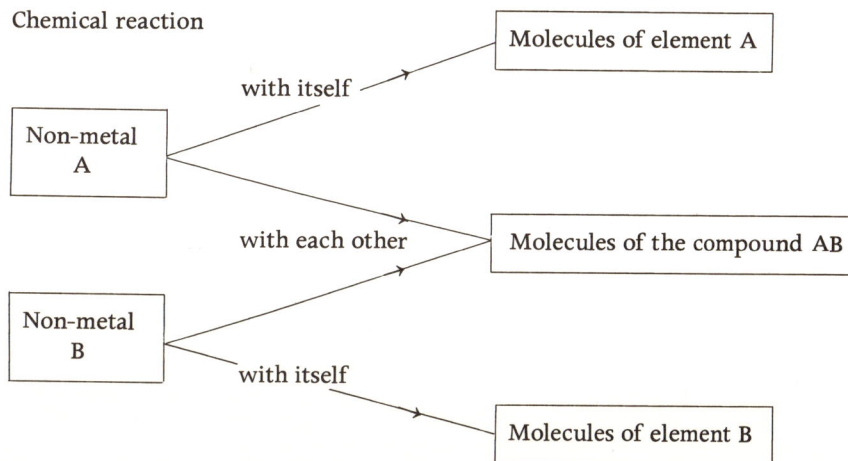

Other everyday molecular compounds are methane (North Sea gas), ammonia, paraffin, petrol and alcohol. Perhaps you can think of some more. Remember, molecular substances are likely to be liquids or gases at room temperature, and will also have the other properties listed at the end of the last Section.

D1.2 The mass of atoms and the mass of molecules

In 1808, a Manchester schoolmaster called John Dalton put forward a theory which explained the behaviour of elements. Dalton's ideas revolutionised chemical thinking and are the basis of modern chemistry. He said that:

a)   elements cannot be simplified by means of chemical reactions because they consist of chemically indivisible particles called atoms

b)   atoms of the *same* element are identical in every way; they have the same mass, size and chemical properties

c)   atoms of *different* elements have different mass, size and chemical properties.

Since each element has its own type of atom with a particular mass which does not change, we should be able to find the mass of the atoms of the different elements by doing experiments. Modern chemists have very accurate methods for doing this, and the mass of each type of atom has now been found. The lightest element of all is hydrogen. All other atoms are heavier than the hydrogen atom. The following list tells you how many times heavier than an atom of hydrogen are the atoms of the other elements listed.

Mass of atoms compared with the mass of a hydrogen atom

Carbon	12
Nitrogen	14
Oxygen	16
Magnesium	24
Sulphur	32
Chlorine	35.5
Calcium	40
Bromine	80
Iodine	127

This means that a carbon atom has 12 times the mass of a hydrogen atom, and that a calcium atom has 40 times the mass of a hydrogen atom. How many times the mass of a hydrogen atom has a sulphur atom? How many times the mass of a hydrogen atom has an oxygen atom?

So we can compare the masses of different atoms, by comparing them all with hydrogen. Can we compare the masses of different *molecules*? Let us try with this example:

A methane molecule is made of 1 carbon and 4 hydrogen atoms, an oxygen molecule is made of 2 oxygen atoms. How much more mass does an oxygen molecule have than a methane molecule?

| methane molecule | 1 carbon atom
+
4 hydrogen atoms | 2 oxygen atoms | oxygen molecule |

We will set about solving the problem by going back and comparing everything with the mass of a hydrogen atom. A carbon atom is 12 times as heavy as a hydrogen atom. So altogether, a methane molecule has $12 + 4 = 16$ times as much mass as one hydrogen atom. An oxygen molecule has $16 + 16 = 32$ times as much mass as one hydrogen atom. So an oxygen molecule has exactly twice the mass of a methane molecule ($2 \times 16 = 32$). We can compare the masses of molecules by comparing the masses of the atoms which make them up with the mass of one atom of hydrogen. We therefore find out how much more mass the molecule has *as a whole* than one atom of hydrogen. We can do this for different molecules and compare them. Try another example:

A sulphur dioxide molecule is made of one atom of sulphur and two atoms of oxygen. How many times the mass of one hydrogen atom does one molecule of sulphur dioxide have? How many times the mass of an oxygen molecule does a sulphur dioxide molecule have?

Remember that the mass of a molecule obtained in this way is, like the masses of the atoms given above, the mass *compared with the mass of one hydrogen atom*. It is not an actual mass, but a *comparative* mass. This method does not tell us how many grams an atom or a molecule weighs, but it does allow us to compare the masses of atoms and molecules.

D1.3 Low mass molecules and high mass molecules

We have seen that, generally, molecular substances are either liquids or gases at room temperature. Sometimes, molecular substances are solids at room temperature. For instance, iodine is a solid. What is it that decides whether a molecular substance will be a liquid, a solid or a gas? Perhaps we can find a rule which holds for all molecular substances.

Here is some information about a group of similar molecular compounds called *alkanes*. These compounds contain only carbon and hydrogen.

Name	Number of atoms of each type per molecule		Mass of molecule compared with mass of 1 atom of hydrogen	Boiling point (°C)
	Carbon	Hydrogen		
Methane	1	4	16 (12 + 4)	−162
Ethane	2	6	30 (24 + 6)	− 89
Propane	3	8	44 (36 + 8)	− 42
Butane	4	10	58 (48 + 10)	− 1
Pentane	5	12	72 (60 + 12)	36
Hexane	6	14	86 (72 + 14)	69

Use the information to plot a graph of the mass of the molecule compared with the mass of one hydrogen atom against the boiling point of the compound. Draw a line for room temperature (15°C):

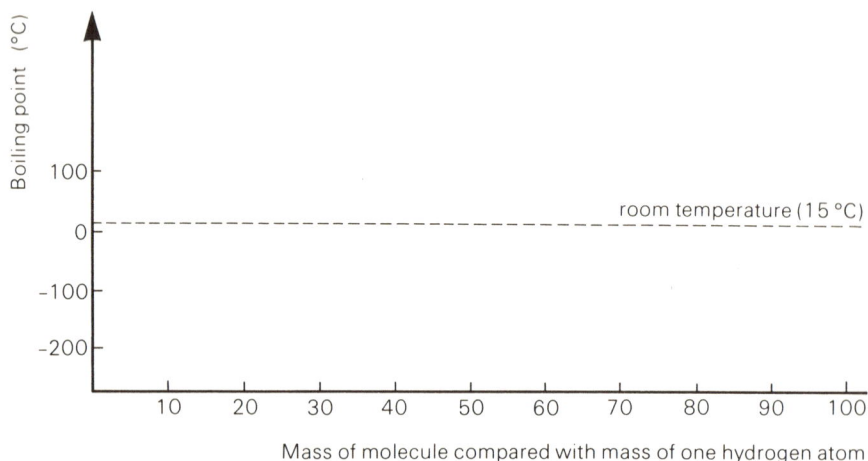

If a compound has a boiling point which is greater than 15°C then it will be a liquid at room temperature. If the compound boils below 15°C it will be a gas at room temperature. Use the graph to decide which of the alkanes are liquids and which are gases at room temperature and write them in a table like this:

Alkanes which are gases at 15°C	Alkanes which are liquids at 15°C

We can easily see that as the mass of the molecule gets greater the boiling point of the compound increases. We could find the melting points for the alkanes in tables – if we did this we would find that melting points also increase as molecules have more mass. Molecules with very high masses have *melting points* above room temperature – they are *solids* under normal conditions. For the alkanes we find that:

Gases	Liquids	Solids
1–4 carbon atoms per molecule	5–15 carbon atoms per molecule	More than 15 carbon atoms per molecule
Low mass molecules		High mass molecules

We seem to have found the rule we were looking for. But what about molecular elements? Does the rule hold for them too? We can find out by looking again at the halogen group of elements.

Halogen	Mass of molecule compared with mass of one hydrogen atom	Gas/solid/liquid
Fluorine (F_2)	38	Gas
Chlorine (Cl_2)	71	Gas
Bromine (Br_2)	160	Liquid/gas
Iodine (I_2)	254	Solid

Clearly the answer is "yes". The greater the mass of the molecule, the higher the melting point and the boiling point of the halogen. We now have a general rule to cover all molecular substances.

For all molecular substances (both elements and compounds) the melting point and boiling point are greater the greater the mass of the molecule.

Notice that it is not how complicated a molecule is (how many different elements make it up) that determines melting point and boiling point. Nor is it the total number of atoms that is important. (Both these points are well illustrated by iodine. It has only one *type* of atom in the molecule and the molecule has a total of only two atoms – but it is a solid nevertheless.) What matters is the total mass of the molecule – the mass of all the atoms added together.

Topic summary and learning

1   Not all molecular substances are compounds. A few elements also exist as molecules.
2   A molecular element is one in which two or more atoms of the same element join together making molecules.
3   Many molecular elements are *diatomic* – they contain two atoms per molecule. Examples are hydrogen (H_2), nitrogen (N_2), oxygen (O_2) and all the halogens.
4   For all molecular substances (both elements and compounds) the melting point and boiling point are greater the greater the mass of the molecule.

5   We can compare the mass of two molecules by comparing each with the mass of one hydrogen atom. If we know the mass of each atom in a molecule compared with the mass of one hydrogen atom, we can add them together and find out how many times heavier than a hydrogen atom the whole molecule is. Two molecules can then be compared and we can see which has the greater mass.

6   The lower the mass of a molecule, the more likely the compound is to be a gas. The greater the mass of a molecule, the more likely the compound is to be a solid.

Learn the information contained in the Topic Summary.

You should remember what we mean by the word *alkane*, and you should remember which alkanes are gases, which liquids, and which solids.

Questions

1   Here is some information about some molecular substances:

	Number of atoms of different elements per molecule			
	Carbon	Hydrogen	Nitrogen	Oxygen
Carbon monoxide	1	–	–	1
Nitrogen	–	–	2	–
Nitrogen dioxide	–	–	1	2

a)   Work out the mass of the molecule of each compound compared to the mass of one atom of hydrogen.

b)   Nitrogen is a gas at room temperature. Do you think that carbon monoxide is also a gas? Explain your answer.

c)   Which molecule has the greater mass – sulphur dioxide or nitrogen dioxide? (see p. 91)

2   Which of the following molecular substances is most likely to be a solid at room temperature?

Substance	A	B	C	D	E
Mass of molecule compared with mass of one atom of hydrogen	2	142	44	32	71

3   Which of the following molecular substances is most likely to be a gas at room temperature?

Substance	A	B	C	D	E
Mass of molecule compared with mass of one atom of hydrogen	46	170	142	32	184

Topic 2
SOME LOW MASS MOLECULES

D2.1   Carbon dioxide

Carbon dioxide is a gas. It is a very important gas because it is involved in the process by which we get our energy from the plants which we eat.

Plants take carbon dioxide, water and energy (which they get from sunlight) to make complicated molecules called sugars, together with oxygen. The process is called *photosynthesis* ("putting together with the help of light").

When we eat a plant, the stored sugars pass into our bloodstream and are transported to our muscles. There they meet oxygen which we have breathed in and the sugars "burn" – and are broken down back into carbon dioxide and water. At the same time, the energy which had been stored in the sugars, and which came originally from the sun, is released. Our muscles use this energy for movement. This reversal of photosynthesis is called *respiration*.

Carbon dioxide is important for other reasons also. It is a heavy gas which puts out flames. So we can use it for smothering fires. The fire-extinguisher in your chemistry laboratory probably contains carbon dioxide. Think how many lives are saved all over the world in one year because of carbon dioxide.

A fire extinguisher producing carbon dioxide

Carbon dioxide also provides the fizz in fizzy drinks, and solid carbon dioxide ("dry ice") is used in some sorts of refrigeration. Carbon dioxide is so important that about 200 000 tons of it are made every year in the United Kingdom alone. The following experiment is best demonstrated by the teacher.

Which elements make up carbon dioxide?

Your teacher will carry out a demonstration.

Experiment 26

The following will be needed: gas jars of carbon dioxide; magnesium ribbon; tongs.

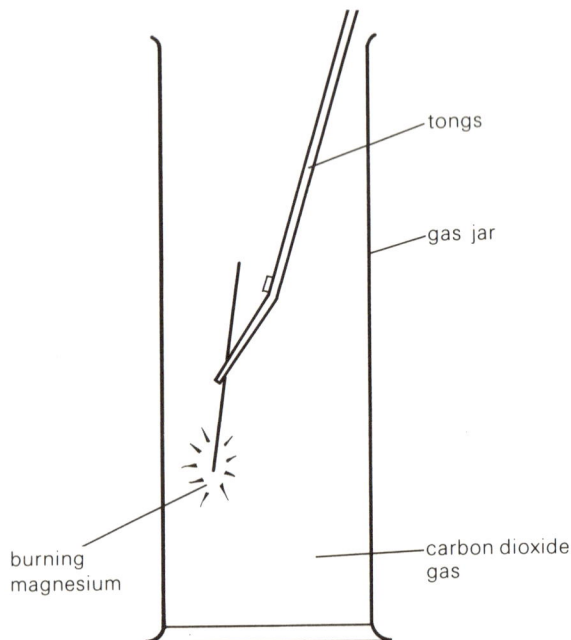

Ignite a piece of magnesium ribbon a few centimetres long in a Bunsen flame and plunge it into a gas jar of carbon dioxide, using tongs. The magnesium will continue to burn for a short time. Repeat using a new piece of magnesium and a fresh gas jar of carbon dioxide. Pass the gas jars round the class for examination.

During the demonstration you should note what happens when the burning magnesium meets the carbon dioxide. When the reaction has finished, look at the inside of the gas jar. You should see that two solids have been produced – a white solid and a black solid.

We know that magnesium burns in air producing a white powder – magnesium oxide. Magnesium oxide is a compound of the elements magnesium and oxygen. The white powder produced when magnesium burns in carbon dioxide is also magnesium oxide. Magnesium has gained the element oxygen from the carbon dioxide. The other product of the reaction was a black solid – the element carbon. So now we know both the products, we can write the word equation:

magnesium (s) + carbon dioxide (g) ⟶ magnesium oxide (s) + carbon (s)

We know from Topic B4 that matter is neither created nor destroyed during a chemical reaction. So the elements present at the end of the reaction must have been there at the beginning.

We can use the word equation to work out:

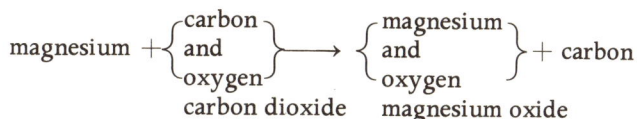

magnesium + {carbon and oxygen} ⟶ {magnesium and oxygen} + carbon

carbon dioxide   magnesium oxide

So carbon dioxide is made of the elements carbon and oxygen.

Where do we get carbon dioxide from?

You will carry out some experiments yourself and will be shown another source of carbon dioxide by your teacher.

You are going to carry out some experiments to find ways of making carbon dioxide. You will try various reactions to see whether they make this gas. So you will need a test which will tell you when you have succeeded.

Carbon dioxide very easily turns limewater (calcium hydroxide solution) "milky". You can show that there is carbon dioxide in the air you breathe out by blowing through a glass tube into some limewater (about 1 cm deep) in a test-tube.

In the reactions which you are going to try now, carbon dioxide (if present) will be made in a test-tube. To test for it, have ready a second test-tube containing limewater 1 cm deep in the position shown:

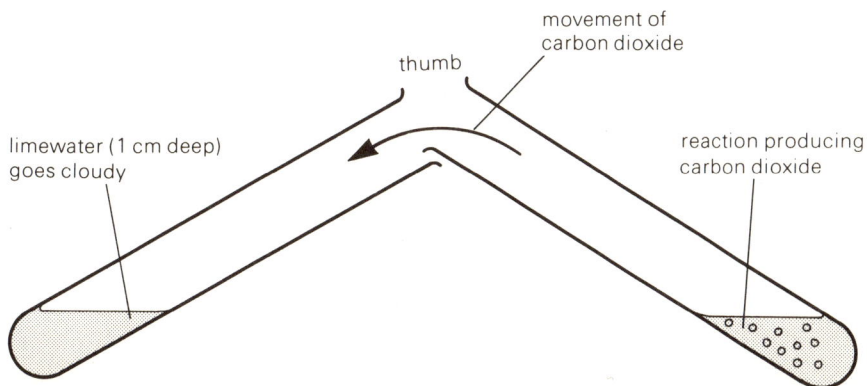

Hold the test-tubes together at an angle (so that you do not tip out their contents). Because carbon dioxide is such a heavy gas, if it is present it can be "poured" like a liquid from one test-tube to another (you can help to keep it in the test-tubes by putting your thumb over the top). Any carbon dioxide made in the experiment will turn the limewater milky.

Always use fresh limewater for each new test.

Experiment 27

The following will be needed: the solids sodium carbonate, copper(II) carbonate, sodium chloride, copper(II) chloride, sodium sulphate and copper(II) sulphate; dilute hydrochloric acid; dilute sulphuric acid.

Take one spatula measure of each of the solids and react it in turn with dilute hydrochloric acid in a test-tube. If a gas is given off, test it to see whether it is carbon dioxide. Draw up a table of your results.

Repeat for each solid using dilute sulphuric acid. Add your results to your table.

You will produce a table like the one below.

Solid	Dilute hydrochloric acid	Dilute sulphuric acid
Sodium carbonate	Carbon dioxide produced	Carbon dioxide produced
Copper(II) carbonate	Carbon dioxide produced	Carbon dioxide produced
Sodium chloride	–	–
Copper(II) chloride	–	–
Sodium sulphate	–	–
Copper(II) sulphate	–	–

The compounds which reacted with the acids to produce carbon dioxide were both carbonates. *Any carbonate will react with any acid to make carbon dioxide.*

The following experiment is best demonstrated by the teacher.

Experiment 28

The following will be needed: filter-pump; Dreschel bottle (or similar); thistle-funnel; candle; watch-glass; sucrose.

A number of carbon-containing compounds can be burned and the resulting gases collected and passed through limewater to show that they contain carbon dioxide, using the apparatus at the top of p. 99.

In each case the limewater will turn milky, showing that carbon dioxide was produced. Most carbon-containing

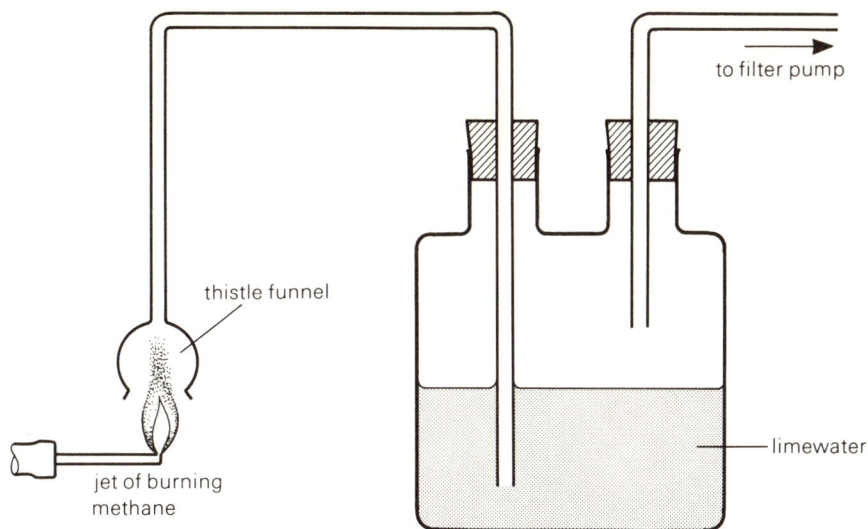

thistle funnel

jet of burning methane

to filter pump

limewater

compounds produce carbon dioxide when they burn. *Compounds which contain only carbon and hydrogen (these compounds are called hydrocarbons) always produce carbon dioxide when they burn.*

Carbon dioxide in the laboratory

We use the reaction between a carbonate and an acid to give us a supply of carbon dioxide in the laboratory. We use the following apparatus:

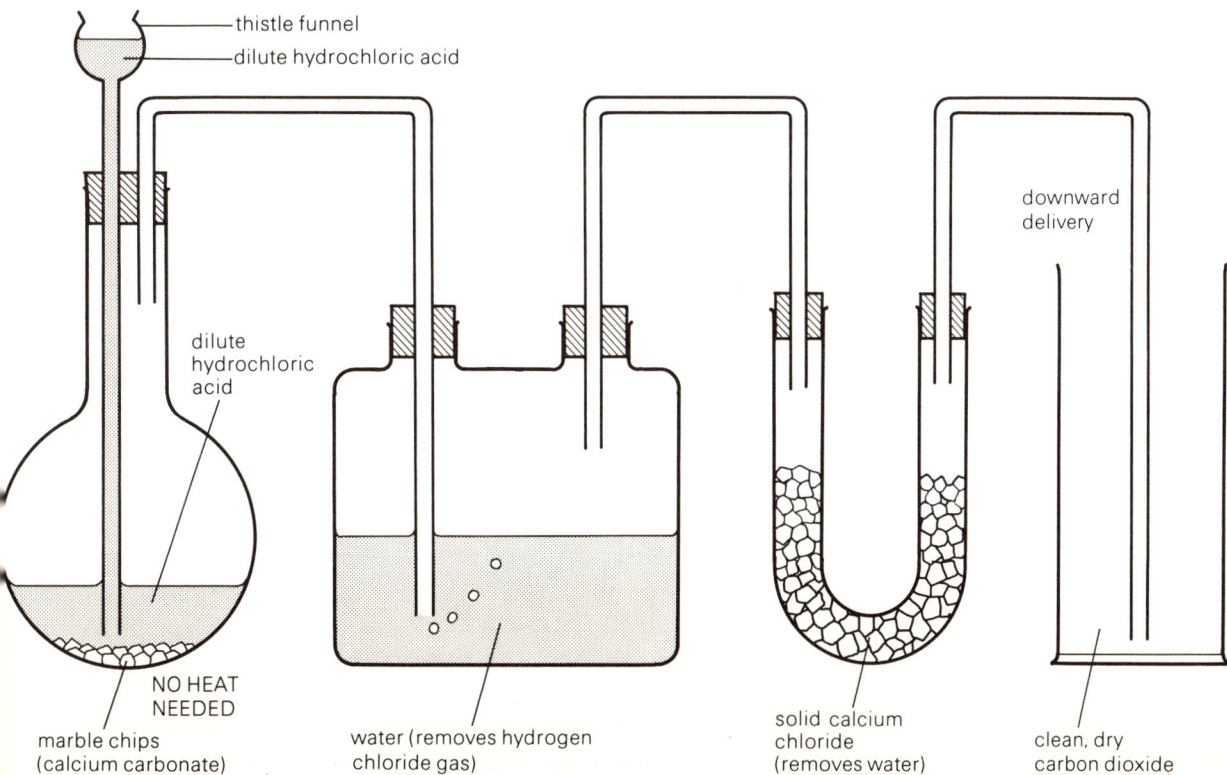

thistle funnel

dilute hydrochloric acid

dilute hydrochloric acid

downward delivery

NO HEAT NEEDED

marble chips (calcium carbonate)

water (removes hydrogen chloride gas)

solid calcium chloride (removes water)

clean, dry carbon dioxide

Your teacher will show you how carbon dioxide is made in the laboratory, and will show you some of its properties.

Dilute hydrochloric acid is reacted with lumps of calcium carbonate (marble) which are often called marble chips. Note that no heat is usually required to make the reaction take place, although heating does speed up the production of carbon dioxide. Note also that *the end of the thistle-funnel must be below the level of liquid in the flask* (Why?)

When the carbon dioxide is first made it is contaminated with hydrogen chloride gas. Fortunately this is very soluble and can be removed simply by passing the gas through water. The carbon dioxide picks up some water during this process and this is then removed using a substance which absorbs water (a drying agent).

Since carbon dioxide is heavier than air it is collected by delivering it downwards into a gas jar.

Your teacher may use gas jars of carbon dioxide which has been collected to demonstrate that it has the following properties:

1   It is heavier than air
2   It extinguishes a burning splint
3   It dissolves in water forming an acid (carbonic acid).

Another property of carbon dioxide is that it will react with white-hot carbon producing carbon monoxide – a very poisonous gas.

D2.2   Water

Water is a liquid at room temperature. Of course, we know it in its solid form (ice) and its gaseous form (steam) as well. We know from Topic B4 (Experiment 22) that water is made from the elements hydrogen and oxygen. They are in the ratio 2 parts of hydrogen: 1 part of oxygen. This gives us the familiar formula for water: H_2O.

Water is a very common substance on Earth (just look at a map of the world and estimate what fraction of the surface is covered with water). Scientists believe that this is one of the major reasons why life evolved on this planet rather than on others such as Venus or Mars, where water is less common.

All life began in the sea. One reason for this is that in the sea the temperature tends to remain fairly constant. On land there is a much greater fluctuation of temperature from day-time to night-time and from summer to winter. Keeping the conditions constant helped life to develop. We still depend on water to keep our environment stable today. If the temperature of the Earth were to drop permanently by 20°C, it is likely that we would eventually run out of fuel to keep ourselves warm – and we would then freeze to death.

Our own bodies are made largely of water, and many of the plants which we eat are almost completely water:

	% by weight of water
Human beings	70
Fish	80
Lettuce	95

But why should living things need to *contain* water? Life is a complicated series of chemical reactions. These need to take place *in solution* – and water is the best substance there is for dissolving things.

Water the solvent

In Topic C2 you carried out an experiment (Experiment 24) which investigated the solubility of different substances in water. Have another look at the Topic and your own notes to remind you of what you did and what you discovered.

The results of Experiment 24 showed that ionic compounds usually dissolve easily in water. Also, some molecular compounds are soluble. In fact, we now know of quite a number of molecular compounds which are soluble in water.

Here is the start of a list of molecular compounds we have met which are soluble in water. Try to finish the list off yourself.

Molecular compounds soluble in water

Ethanol
Carbon dioxide
Hydrogen chloride
Ammonia

So water is an excellent solvent. It will dissolve most ionic and many molecular compounds. It provides a stable environment in which the chemical processes of life can take place. It is also vital to our everyday life.

Water in everyday life and in industry

Each of us uses about 30 litres of water every day. We use it for drinking, for washing ourselves, for washing our clothes, for flushing toilets, for watering household plants. We use it for watering the garden, for cleaning the windows and washing cars. We make tea, coffee, wine and beer using water; we use it for making ice-cream. We cook our food in water, and then we use water for washing-up the pots and pans, dishes and cups and saucers afterwards. You can probably think of many other everyday uses of water – the list is almost endless.

In addition to the 30 litres we use ourselves, industry uses another 30 litres every day for each one of us. As well as its common uses as a solvent for chemical processes carried out on a large scale, water is also used in industry in dyeing, in steam turbines and as a means for cooling reaction vessels down. For each of these purposes, the water has to be *pure*. But water is such a good solvent that it nearly always contains dissolved substances which have to be removed. Where does industry get its water from, and how is it purified?

Sources of water

When water falls as rain it is almost pure. But even then it contains nitrogen, oxygen and carbon dioxide which have dissolved as the raindrops passed through the atmosphere. If the rain can be collected soon after falling, reasonably pure water can be obtained. So a reservoir high up in the mountains contains almost pure water, because nothing else has had a chance to dissolve in it. This is particularly true if the mountains are made of a very hard rock like granite which will not dissolve.

But if the water is not collected straight away and is allowed instead to run down the mountainside, it will dissolve solids from the soil it passes through. These will include substances such as calcium and magnesium carbonate, calcium and magnesium hydrogencarbonate, sodium chloride, various nitrates and some ammonium salts. Organic matter will also dissolve and bacteria and other microscopic organisms will be picked up. By the time the water reaches a major river, it may contain quite a lot of these substances, depending on the types of rock which the water has passed over since falling as rain. In different regions, therefore, the water will contain different dissolved substances.

Eventually, all water passes down to the sea, where all the dissolved substances in river water have been accumulating over millions of years. It is not surprising, therefore, to find that seawater is quite impure – you only need to taste it!

Rain water ⟶ river water ⟶ sea water
Increasingly impure water
⟶

Because of the quantities of water which are required for use in the house and in industry, very little of it can be obtained in a pure state. Most water is taken from rivers, or from reservoirs into which rivers flow, and it therefore contains many dissolved substances. These impurities can cause major problems in industry if they are not removed. For instance, many dissolved substances cause dyes to change their colour. Where water is heated (for instance, in a steel boiler or in a domestic kettle) the concentration of dissolved substances builds up until a solid forms. This solid can cause blockages, and may even cause boilers to burst.

So before water is used, it has to have the dissolved

substances removed. The most obvious way of purifying water is to distil it. When water boils, the steam comes off and leaves the dissolved substances behind. So if the steam is then condensed it produces pure water. We need pure water in a laboratory, and we often get it by distillation. Your chemistry laboratory may have a still which your teacher can show you. Where else do you find pure water which has been produced in this way?

Unfortunately, distillation is very expensive because of the energy required to boil the water. It would be too costly to purify the huge quantities of water which industry needs using distillation. So less costly methods have to be devised.

A troublesome impurity – hardness

You will do experiments which show the causes and different types of hardness.

Hard water is water which will not easily form a lather with soap.

If you live in an area where the water is hard, you will be familiar with what happens when you try to lather water. It takes a lot of soap, and an unpleasant "scum" is formed at the same time. Hardness is due to dissolved substances. We will now try to find out what those substances are.

Experiment 29

The following will be needed: solutions of calcium hydrogencarbonate, calcium nitrate, calcium sulphate, sodium hydrogencarbonate, sodium nitrate, sodium sulphate, distilled water; soap solution.

Take some distilled water and find out how hard it is by seeing how many drops of soap solution are needed to cause distilled water 2 cm deep in a test-tube to lather. Add one drop of soap solution from a teat-pipette and shake it vigorously by placing your thumb over the end of the test-tube. Then put the tube in a rack, and wait for 20 seconds. If a lather remains after this time, we can say that one drop of soap solution was needed. If the lather does not remain, add another drop of soap solution and shake again. Wait for a further 20 seconds. Carry on until the lather remains, and record how many drops of soap solution were needed.

Take three of the other solutions *at random* and compare them for hardness with the distilled water by repeating the above procedure. Use a depth of 2 cm of distilled water in a test-tube, and find how many drops of soap solution are required to form a lather which lasts for 20 seconds. Record your result for each solution. Other groups will be using different solutions and a complete picture will be built up by your teacher at the end of the lesson.

Now take the three other solutions which you have *not* already used, and put each of them in a test-tube to a depth of 2 cm. Place the three test-tubes in a beaker one-third full of water, and heat this on a tripod and gauze until it boils. Continue the boiling for 10 minutes and remove the test-tubes to a rack. Allow them to cool and note any changes which have taken place. Then test the three solutions for hardness as above, recording your result for each solution.

By the time the whole class has finished, each of the six solutions will have been tested for hardness before and after boiling.

Using the results from this experiment, we can call each solution "soft" or "hard" depending on how many drops of soap solution were needed before a lather formed.

Solution	Before boiling	After boiling
Distilled water	Soft	Soft
Calcium hydrogencarbonate	Hard	Soft
Calcium nitrate	Hard	Hard
Calcium sulphate	Hard	Hard
Sodium hydrogencarbonate	Soft	Soft
Sodium nitrate	Soft	Soft
Sodium sulphate	Soft	Soft

A solution of calcium hydrogencarbonate contains calcium cations and hydrogencarbonate anions. A solution of sodium hydrogencarbonate contains sodium cations and hydrogen-carbonate anions. *Both solutions contain hydrogencarbonate ions, but only the calcium hydrogencarbonate was hard* (before boiling). So the hardness must be due to the calcium cations. Both calcium nitrate solution and calcium sulphate solution contain calcium cations, so they are both hard. Why do calcium ions cause hardness?

The chemical name for soap is sodium stearate. It is the stearate anions in solution which cause the lather to form. When calcium ions are present, however, a reaction takes place.

Hard water   soap
Calcium ions (aq.) + sodium stearate (aq.)

                                      scum
      $\longrightarrow$ calcium stearate (s) + sodium ions (aq.)

The calcium stearate produced is insoluble. The result of the reaction is therefore that the stearate anions are *no longer in the solution* and no lather is formed. The calcium stearate (scum) contains all the stearate ions from the soap.

If enough soap is added, all the calcium ions can be used up (*taken out of the solution*) in producing scum. If more soap is added after this has happened, a lather will form as though the water were soft. So the amount of soap needed to soften water is a measure of how much hardness was there originally. The more calcium ions in a solution, the more soap is needed to get rid of them as scum. So now we can see what causes hardness (calcium ions in solution), and why hard water uses up a lot of soap and causes a scum to form. But there is one result from the last experiment which we have not explained yet. Why did calcium hydrogencarbonate solution become soft after boiling?

Calcium hydrogen carbonate is unstable when heated. It breaks down, forming insoluble calcium carbonate. Because calcium carbonate is a solid, the calcium ions are *no longer in the solution*, so the water is no longer hard.

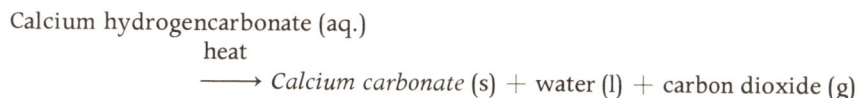

Calcium hydrogencarbonate (aq.)

$$\xrightarrow{\text{heat}} \textit{Calcium carbonate} \text{ (s)} + \text{water (l)} + \text{carbon dioxide (g)}$$

You may have seen some calcium carbonate forming as a white precipitate if you were one of those who boiled calcium hydrogencarbonate solution.

Water which can be softened by boiling is called temporary hard water. Temporary hardness is caused by dissolved calcium hydrogencarbonate.

Water which cannot be softened by boiling is called permanent hard water. Permanent hardness is caused by dissolved calcium salts which are not calcium hydrogencarbonate (often calcium sulphate).

The hardness of the water used in this kettle has caused a precipitate of calcium carbonate "scale"

Man-made impurities: pollution

When a chemical plant uses water, care must be taken about the impurities, such as hardness, already in the water. But it is also important to be careful about the impurities left behind in the water by the chemical process. We normally call these impurities *pollution*.

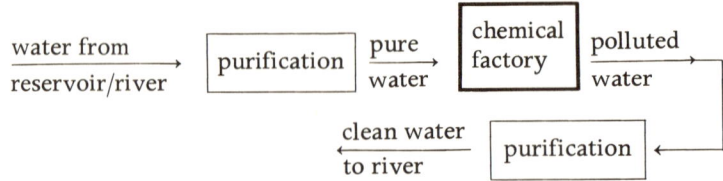

water from reservoir/river → | purification | → pure water → | **chemical factory** | polluted water →

clean water to river ← | purification | ←

The water has to be purified before it can be returned to the river. In a similar way, the water we use ourselves cannot simply be left as it is and returned to a river. It too has to be treated.

Industrial pollution

Type of pollution	Treatment
Heat	Allow to cool
Solid matter in suspension	Filter and then leave to settle
{ Dissolved solids, liquids and gases	Treat chemically

Domestic pollution

Type of pollution	Treatment
Sewage	Pass through filtration beds and allow to settle. Break down with bacteria.
Detergents	Treat chemically.

This river has been heavily polluted by detergent foam

Often, water is used in a chemical plant for cooling down a chemical reaction. The water takes away the heat, and may be raised nearly to boiling point. Obviously, if water at this temperature were simply pumped back into a river, all the fish would be killed. Similarly, dissolved chemicals are often poisonous to fishes and other forms of life and have to be removed.

To prevent pollution of the rivers and seas, most towns have their own Sewage Works where fouled domestic water is made harmless before it is allowed to leave.

Topic summary and learning

1   Carbon dioxide is a molecular compound of the elements carbon and oxygen. Since the molecule has low mass, carbon dioxide is a gas at room temperature.

2   Carbon dioxide is involved in the process by which green plants store the sun's energy. It has many other important uses.

3   Carbon dioxide turns limewater "milky".

4   Any carbonate will react with any acid to make carbon dioxide.

5   Hydrocarbons are compounds which contain only the elements carbon and hydrogen. They all produce carbon dioxide when they burn.

6   Carbon dioxide is made in the laboratory by reacting calcium carbonate (in the form of marble chips) with dilute hydrochloric acid.

7   Carbon dioxide is heavier than air, it extinguishes flames and it dissolves in water forming an acid. It reacts with white-hot carbon producing the poisonous gas carbon monoxide.

8   Water is a compound of the elements hydrogen and oxygen. The water molecule has low mass and so water is a liquid at room temperature.

9   Water is essential to life.

10   Water is an excellent solvent. Most ionic and some molecular compounds are soluble in water.

11   Water has many uses in the home and in industry, and is used in large quantities.

12   Water from different sources contains different dissolved substances. Hard water is water which will not easily lather, and it is caused by the presence of calcium ions.

13   Dissolved calcium hydrogencarbonate causes temporary hardness, which can be removed by boiling.

Dissolved calcium sulphate causes permanent hardness, which cannot be removed by boiling.

14   Man-made pollution of water takes various forms, and has to be cured in a variety of ways.

You should learn the information contained in the Topic Summary.

You should remember the different uses of carbon dioxide, the nature of photosynthesis and respiration, and the details of the laboratory preparation of carbon dioxide.

You should remember the different ways in which water is important to life, the names of molecular compounds soluble in water, the different uses we make of water, and the problems caused by dissolved substances in the water we want to use. You should also remember the word-equation for the reaction of hard water with soap, and the nature and methods of removal of man-made pollution of water.

Questions

1   Carbon dioxide dissolves in water to form an acid. Why would you expect this to be true? Which other elements have oxides which dissolve in water to form acids?

2   Write a word-equation for each of the following processes:

a) photosynthesis

b) respiration

For each process, state whether it is exothermic or endothermic.

3   The "milkiness" produced when carbon dioxide is passed into calcium hydroxide solution is due to the insoluble calcium carbonate which is formed. Water is also formed in this reaction. Write a word-equation.

Write a word-equation for the reaction which takes place when carbon dioxide is passed into a solution of sodium hydroxide. Why do you think no "milkiness" is produced in this case?

4   If the "milkiness" produced by passing carbon dioxide into limewater were filtered off, a white solid would be obtained. Describe what you would see if dilute hydrochloric acid were added to this white solid.

5   Here are some word equations for the reactions of some carbon-containing compounds:

Calcium hydrogencarbonate (aq.)

$\xrightarrow{\text{heat}}$ calcium carbonate (s) + carbon dioxide (g) + water (l)

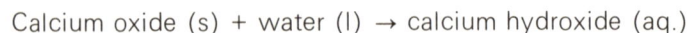

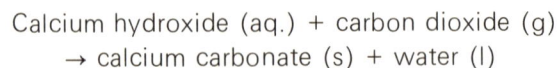

Calcium carbonate (s) + carbon dioxide (g) + water

$\xrightarrow[\text{pass } CO_2]{\text{cold}}$ calcium hydrogencarbonate (aq.)

Calcium carbonate (s) $\xrightarrow{\text{heat}}$ calcium oxide (s) + carbon dioxide (g)

Calcium oxide (s) + water (l) → calcium hydroxide (aq.)

Calcium hydroxide (aq.) + carbon dioxide (g)

→ calcium carbonate (s) + water (l)

Use the word-equations to complete the following diagram, which shows how these compounds are related to each other, by writing in the names of the missing compounds

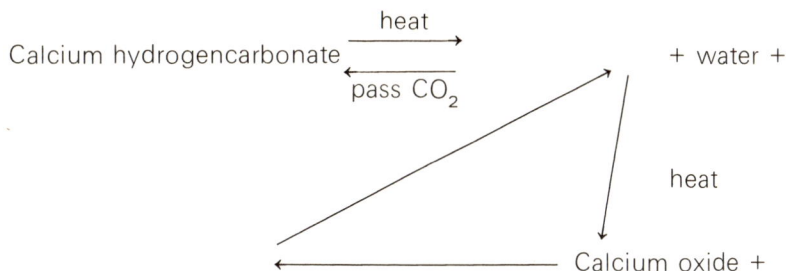

Calcium hydrogencarbonate $\xrightarrow[\text{pass } CO_2]{\text{heat}}$ + water +

heat

Calcium oxide +

When carbon dioxide is passed into limewater, a white cloudiness forms. What would you see if carbon dioxide were passed into the solution after the milkiness had formed? Give another name for the solution which is formed. What would be the effect (state what you would see, and name any substances formed) if soap solution were added to this solution and shaken?

6 Explain (in your own words as far as possible:

a) the difference between temporary hardness and permanent hardness

b) the problems caused by man-made pollution of water.

Topic 3
SOME HIGH MASS MOLECULES

D3.1 Carbon compounds

Look at the Periodic Table on p. 22, and find the element carbon. You will see that it is in group 4 – right in the middle of the Table. You will also see from the Table that carbon is a non-metal.

But carbon is a special non-metal. It is particularly good at reacting with other non-metals. It rarely reacts with metals. So almost all its compounds are molecular.

There is something else which is special about carbon. As well as being good at reacting with other non-metals, it is good at reacting with *itself*. Molecular compounds therefore exist which have many carbon atoms in them. You have already met some of these compounds – the *alkanes*.

In these molecules with many carbon atoms, the carbon atoms are joined together in long chains:

$$-\overset{|}{\underset{|}{C}}-\overset{|}{\underset{|}{C}}-\overset{|}{\underset{|}{C}}-\overset{|}{\underset{|}{C}}-\overset{|}{\underset{|}{C}}- \qquad \text{a carbon chain.}$$

Sometimes they are joined in rings of six atoms:

a six-membered carbon ring.

For instance, a molecule of the alkane butane looks something like:

$$H-\overset{\overset{\displaystyle H}{|}}{\underset{\underset{\displaystyle H}{|}}{C}}-\overset{\overset{\displaystyle H}{|}}{\underset{\underset{\displaystyle H}{|}}{C}}-\overset{\overset{\displaystyle H}{|}}{\underset{\underset{\displaystyle H}{|}}{C}}-\overset{\overset{\displaystyle H}{|}}{\underset{\underset{\displaystyle H}{|}}{C}}-H$$

The butane molecule has four carbon atoms joined together in a chain. In some compounds there are rings and chains in the same molecule. An example might be:

The arrangement of joined-together carbon atom acts as a

110

skeleton for the molecule. Many different atoms are able to join on to this skeleton – including of course more carbon atoms. So very high mass molecules (having many carbon and hydrogen atoms) are possible.

But because different atoms may join on to the same skeleton, many different compounds can result from the same skeleton of carbon atoms (if the molecule is different, the compound is different). For example:

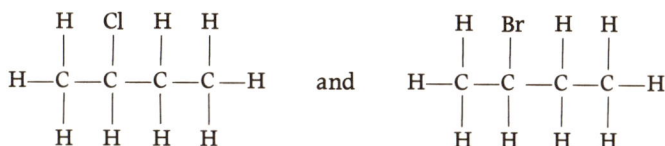

$$
\begin{array}{cccc}
\text{H} & \text{Cl} & \text{H} & \text{H} \\
| & | & | & | \\
\text{H}-\text{C}-\text{C}-\text{C}-\text{C}-\text{H} \\
| & | & | & | \\
\text{H} & \text{H} & \text{H} & \text{H}
\end{array}
\quad \text{and} \quad
\begin{array}{cccc}
\text{H} & \text{Br} & \text{H} & \text{H} \\
| & | & | & | \\
\text{H}-\text{C}-\text{C}-\text{C}-\text{C}-\text{H} \\
| & | & | & | \\
\text{H} & \text{H} & \text{H} & \text{H}
\end{array}
$$

are different compounds based on the same carbon skeleton. Since there are a large number of different carbon skeletons which are possible, and since each skeleton can lead to many different compounds, the total number of compounds of carbon which are possible is enormous. In fact, there are an infinite number of carbon compounds which can be made.

Of all the compounds of all the elements which have been made up to the present time, there are more compounds of carbon than of all the other elements put together.
Starting with a framework of two carbon atoms:

$$
\begin{array}{cc}
| & | \\
-\text{C}-\text{C}- \\
| & |
\end{array}
$$

write down the molecules which are possible if only two other atoms ("A" and "B") are used. Here is one of them:

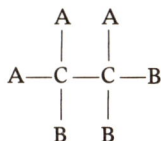

$$
\begin{array}{cc}
\text{A} & \text{A} \\
| & | \\
\text{A}-\text{C}-\text{C}-\text{B} \\
| & | \\
\text{B} & \text{B}
\end{array}
$$

See how many others you can find (you do not always need to have three "A"s and three "B"s).

Try the same thing with a framework of three carbon atoms.

Because there are so many compounds of carbon, a whole branch of chemistry is devoted to their study. This part of chemistry is called *organic* chemistry.

Some carbon compounds have low mass molecules. Carbon dioxide and the alkane methane are examples of carbon compounds which are gases at room temperature. But many of the compounds of carbon have high mass molecules. These solids are often very important in everyday life.

D3.2 Giant molecules

Making giant molecules

Giant molecules can be made by making low mass molecules join together. In this way, a very long carbon skeleton can be produced. Suppose we start with low mass molecules having a skeleton of two carbon atoms. If enough of these are joined together, we can rapidly build up a long carbon chain. For example:

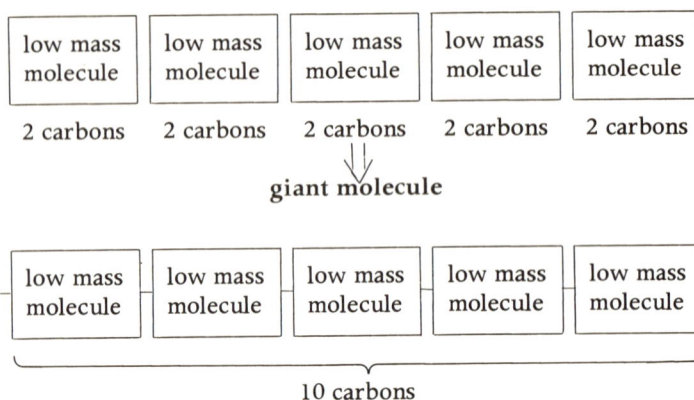

low mass molecule	low mass molecule	low mass molecule	low mass molecule	low mass molecule
2 carbons	2 carbons	2 carbons	2 carbons	2 carbons

giant molecule

low mass molecule	low mass molecule	low mass molecule	low mass molecule	low mass molecule

10 carbons

This process can continue until the molecule contains *thousands* of atoms. In this case we call the molecule a giant molecule. The low mass molecules are *units* which go together to make up the giant molecule. They are like the bricks which are the units in a brick wall.

The individual units which go together to make up the giant molecule are called *monomers* ("mono" means "one"). The giant molecule itself is called a *polymer* ("poly" means "many"). The process of making the giant molecule from many low mass molecules is called *polymerisation*.

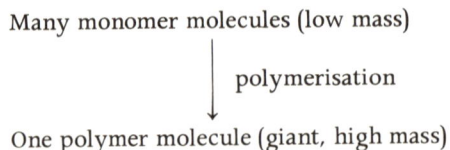

Many monomer molecules (low mass)

polymerisation

One polymer molecule (giant, high mass)

Your teacher may be able to show you some models which will help you to understand the process of polymerisation.

Giant molecules in nature

Nature often makes polymers. One reason for doing this is to "store" the monomer molecules. We saw in the last Topic that green plants make sugars out of carbon dioxide, water, and the sun's energy. Once these sugars have been made, they have to be stored somewhere until they are needed. One way in which plants store sugars is to make giant polymers out of them. The polymers can be easily stored.

Many green plants store sugars in the form of *starch*. Starch is a polymer made up from very many sugar monomers.

This diagram shows part of the natural polymer, starch:

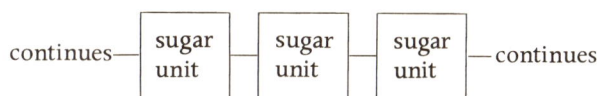

continues— | sugar unit | — | sugar unit | — | sugar unit | —continues

Starch is found in potatoes, in rice, and in many other crops. When the plant wants to use the sugar, it can break down the starch polymer and obtain the sugar. A single starch molecule usually contains many *tens of thousands of sugar units* – a giant molecule indeed!

Man-made giant molecules

You will make the man-made polymer "nylon".

Scientists have discovered that many low mass molecules with a carbon skeleton can be made to form polymers under the right conditions. These man-made polymers are called *plastics*. Different plastics have different properties, and it is now possible to make plastics "to order" for a given purpose. However, most have a number of properties in common:

1   Since they have giant molecules, plastics exist as solids at room temperature.
2   Heating causes plastics to liquefy. So plastics can be poured into moulds when hot and allowed to solidify, giving the required shape.
3   Many plastics are strong and light and are therefore useful for making a wide range of objects.
4   Plastics are waterproof.
5   Many plastics are very cheap to make.
6   Many man-made polymers can be turned into fibres for making clothes.

As with any other chemical process, plastics can be made cheaply only if the raw materials can be obtained at a low cost. The monomers used for making plastics are all obtained (directly or indirectly) from petroleum ("crude oil").

Petroleum is a mixture of many carbon compounds, ranging from low mass molecules (gases) to high mass molecules (solids). The different components in crude oil can be separated by *fractional distillation*. You may have carried out the fractional distillation of crude oil in the past. If not, and if there is enough time, you may be allowed to carry this out now.

Experiment 30

The following will be needed: a sample of crude oil; test-tube with side-arm; thermometer (0°–360°C) in bung for test-tube; delivery tube and rubber connector; Rocksill wool.

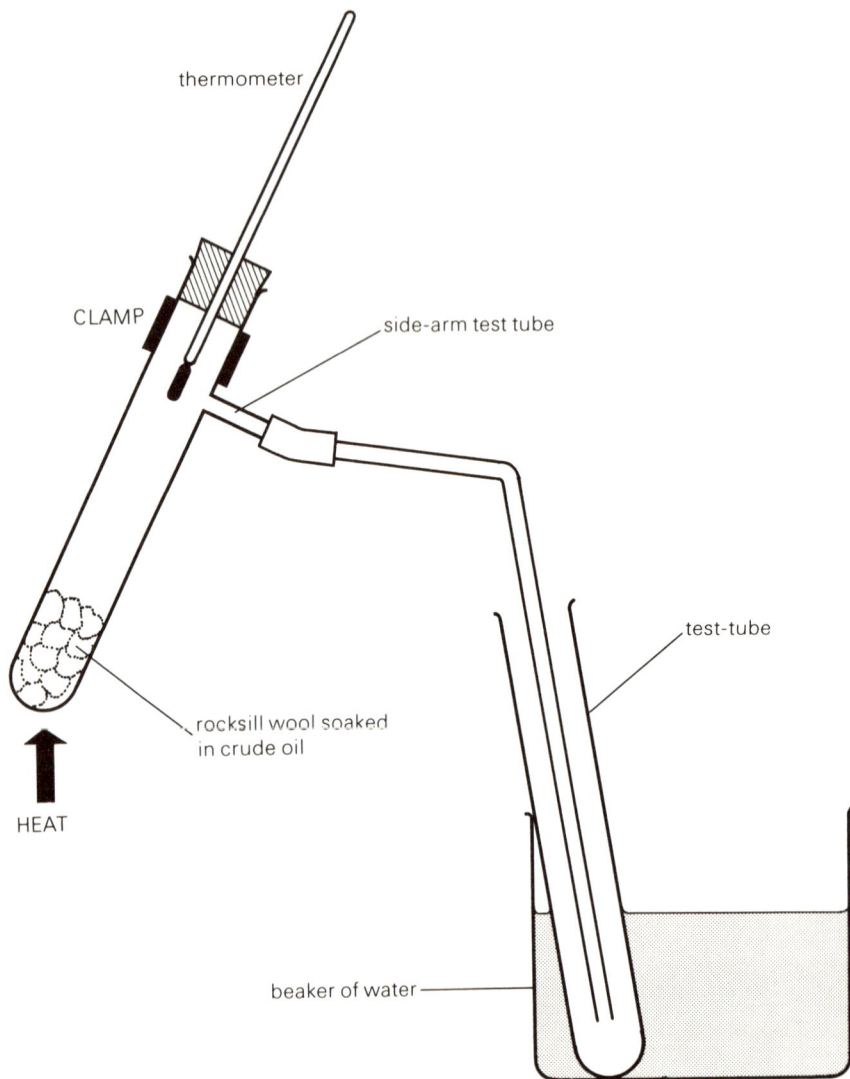

thermometer

CLAMP

side-arm test tube

test-tube

rocksill wool soaked in crude oil

HEAT

beaker of water

Roll the Rocksill wool into a ball and push it to the bottom of the side-arm test-tube with a glass rod. Carefully add enough crude oil to saturate the Rocksill wool. Add this·a little at a time, and prod the Rocksill wool with the glass rod to help the crude oil soak into it. Pour away any crude oil which will not soak into it.

Attach the delivery tube to the side-arm. Clamp the apparatus above the bench so that a test-tube in a beaker of cold water can be placed beneath the delivery tube.

Put the thermometer in its bung into the mouth of the test-tube.

Start to heat the crude oil, watching the temperature recorded on the thermometer. When the temperature reaches 100°C, remove the collecting test-tube to a test-tube rack, and replace it with another.

Continue heating, and collect in the new test-tube the mixture which distils over between 100°C and 150°C.

When the temperature reaches 150°C replace the collecting test-tube and collect the mixture which boils between 150°C and 200°C.

Stop heating. You will see that a solid remains in the side-arm test-tube.

You have split the crude oil into four "fractions".

1 room temperature – 100°C
2 100°C–150°C
3 150°C–200°C
4 remainder (boils above 200°C).

As the boiling points of the compounds in the mixture increase (as the molecules have higher mass) you will see that the colour gets darker, and the mixture becomes less "runny". A fractional distillation of crude oil in industry would give the following fractions:

Fraction	Range of boiling point	Use
Gas	Below 40°C	Methane, camping gas
Petrol	40°C–180°C	Motor fuel, aeroplane fuel
Kerosene	180°C–300°C	Paraffin for lighting and heating
Heavy oils	Above 300°C	Diesel oil, lubricating oils
Wax		Petroleum jelly, paraffin wax
Asphalt		For roadmaking

An oil refinery, where the fractional distillation of crude oil is carried out on a large scale

Since many fractions of crude oil can be used for fuels of different types, it is an important *energy resource* as well as an important source of molecules for making man-made polymers. The following table gives the amounts of crude oil present in different parts of the world, and the amounts which are used in different places.

Reserves (estimated 1979)	Millions of tonnes
OPEC (mainly Kuwait, Saudi Arabia, Iran, Iraq)	3800
USA	4210
USSR	682
UK	200

Consumption (1979)	Millions of tonnes per year
USA and Canada	850
Western Europe	700
Japan	230

Which part of the world has the biggest surplus of oil over and above what it uses for its own needs?

Which part of the world must buy the most oil from other countries to satisfy its needs?

Which part of the world uses the highest proportion of the world's reserves of oil every year?

Which part of the world comes the nearest to being able to supply itself with oil without having a surplus?

Drilling for oil beneath the North Sea takes place from platforms like this

When the oil-producing countries of the world increase the price of oil, the price of many household goods made from plastic also goes up, since the plastic automatically becomes more expensive to make. The following table gives some of the common plastics used every day and whose monomers are obtained from oil.

Monomer	Plastic	Uses
Ethene	Polyethene ("polythene")	Polythene bags, soft plastic toys
Vinylchloride	Polyvinylchloride (P.V.C.)	Plastic rainwear, pipes
Styrene	Polystyrene	"Expanded" polystyrene used for packaging

Sometimes, polymers are made from two monomers. The monomers take up alternate positions in the polymer molecule:

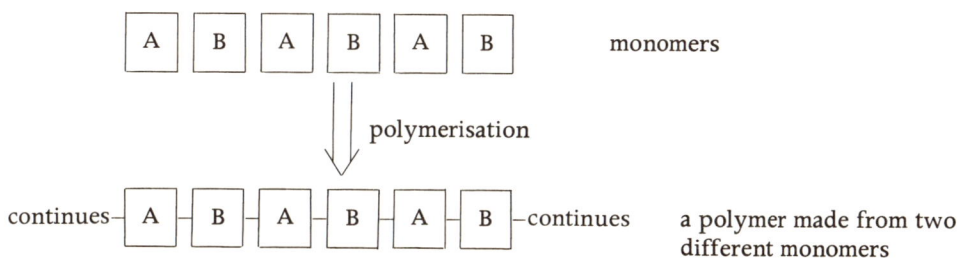

| A | B | A | B | A | B | monomers

polymerisation

continues— A — B — A — B — A — B —continues   a polymer made from two different monomers

An important example of this is nylon.

Experiment 31

The following will be needed: 5% solutions of hexamethylene diamine (in water) and adipyl chloride (in tetrachloromethane); 10 cm³ beaker (or crucible); tongs; glass rod.

Place about 4 cm³ of the solution marked "adipyl chloride" in the small beaker or crucible provided. Carefully add the same volume of the solution marked "hexamethylene diamine". If you are careful, the two solutions will not mix, and you will be able to see a "skin" where they meet.

Using a pair of tongs, pull out some of this skin and wind it onto a glass rod. Continue to pull out more of the fibre by turning the glass rod.

You should be able to make a fibre a few metres long. This fibre is nylon.

The two monomers in nylon are adipyl chloride and hexamethylene diamine. They form the polymer in the following way:

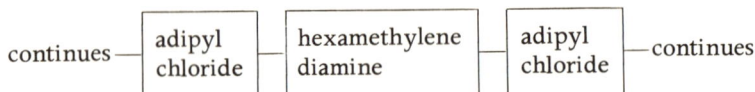

continues — | adipyl chloride | — | hexamethylene diamine | — | adipyl chloride | — continues

Like starch, nylon is truly a giant molecule. There are usually thousands of adipyl chloride and hexamethylene diamine units in a single molecule of nylon.

Topic summary and learning

1   Carbon atoms can join together to form chains and rings. These form the skeletons for a vast number of compounds, some of which have very high mass molecules indeed.

2   There are more compounds of carbon than of all the other elements put together.

3   Giant molecules (*polymers*) can be made by joining together low mass molecules (*monomers*) which have a carbon skeleton. This process is called *polymerisation*.

4   An important natural polymer is starch. The monomer units which go together to make up starch are the sugars produced in green plants by photosynthesis.

5   Man-made polymers are called plastics. The monomers for making plastics are obtained from crude oil.

6   Many of the fractions obtained from crude oil are fuels, making it an important energy resource.

7   Plastics have many useful properties, and some can be made into fibres for the production of clothing. Nylon is a polymer which is made from *two* monomer molecules.

Learn the information contained in the Topic Summary.

You should also be able to explain how polymers are formed from (a) one and (b) two monomers. You should know why polymers have different properties from the monomers from which they are formed.

You should remember the structure of starch, and the properties which make plastics so useful.

Questions

1   How many different compounds can be made by joining up the two elements "A" and "B" in different combinations to the three vacant places in the skeleton:

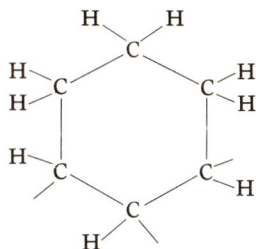

2   Starch is an energy store. In what form is the energy stored, and how does a plant obtain this energy?

3   Make a list of *ten* everyday objects and the man-made plastics which they contain. Why did pop records have to wait for the invention of plastics? Which property of plastics allows us to make things like records?

4   One of the causes of "inflation" is the rise in prices of goods in the shops. Why can inflation sometimes be caused by an increase in the price of crude oil?

5   Which part of the world would probably suffer most by an increase in the price of crude oil? Give a reason for your answer. Why was the discovery of oil in the North Sea so important for this country?

6   Which special property of carbon atoms allows giant molecules to be formed?

EXPLAINING MOLECULES

A molecule is made up of two or more atoms combined together by covalent bonds. Each molecule contains atoms of all the elements which are in the compound. The atoms of the different elements are in the same ratio in the molecule as they are in the compound overall. So, each molecule is the chemical compound – it is the smallest particle of the compound which can exist.

Suppose we take a collection of molecules of a compound like methane:

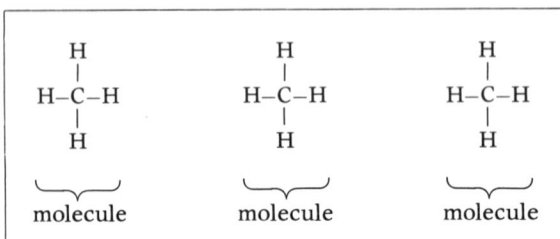

```
      H              H              H
      |              |              |
  H–C–H          H–C–H          H–C–H
      |              |              |
      H              H              H

   ⌣‿⌣            ⌣‿⌣            ⌣‿⌣
  molecule        molecule        molecule
```

If we separate the molecules from each other, but do not alter the molecules themselves, we will still have the same chemical compound:

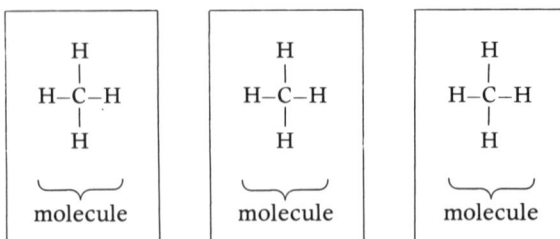

```
      H              H              H
      |              |              |
  H–C–H          H–C–H          H–C–H
      |              |              |
      H              H              H

   ⌣‿⌣            ⌣‿⌣            ⌣‿⌣
  molecule        molecule        molecule
```

It is fairly easy to do this, because the forces between molecules are weak. When we increase the distance between molecules in this way, we might, for instance, change the substance from a solid to a liquid, or from a liquid to a gas. Because we do not need to supply much energy to do this (to overcome the weak forces between the molecules), it can happen at a low temperature. So molecular substances will melt and boil at low temperatures.

But we know that the higher the mass of a molecule, the higher will be the melting point and boiling point of the substance. Why should this be? Temperature is a measure of the energy which the molecules have. And energy is dependent upon both the mass of the molecule and the speed with which it moves. So, at the same temperature, molecules which have different masses will be moving at different speeds. The higher the mass, the lower the speed. The lower the mass, the higher the speed.

Hydrogen   Methane   Butane

molecules have higher mass
————————————————————————→
molecules move at greater speed
←————————————————————————

So at 25°C, for instance, molecules of hydrogen will be moving at a higher speed than molecules of methane. The lower the mass of a molecule, the more rapidly it will be moving, at a given temperature. When a liquid boils, molecules escape from the surface of the liquid into the gas. The faster the molecules are moving in the liquid, the more likely they are to be able to escape. When all the molecules are moving fast enough to escape, the liquid boils. So the boiling point of a liquid is the temperature at which all the molecules are moving fast enough to escape. Suppose we compare two molecular substances, A (low mass molecules) and B (high mass molecules):

Temperature (°C)	% of molecules moving fast enough to escape into gas	
	A	B
10	60	20
25	100	50
40	100	80
55	100	100

Substance A boils at a lower temperature (has a lower boiling point) than substance B. Substances with low mass molecules have lower boiling points than substances with high mass molecules. For the same reason, they also have lower melting points than substances with high mass molecules.

We have seen that the forces between molecules are weak. But what about the forces within molecules? How strong are the forces which hold the atoms in a molecule together? Suppose we look at what happens when a molecule such as methane takes part in a chemical reaction. A new chemical will be produced

at the end of the reaction. This means that a new molecule will
be produced.

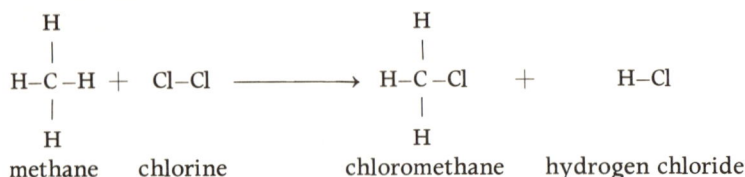

$$
\underset{\substack{\text{methane}}}{H-\overset{\displaystyle H}{\underset{\displaystyle H}{\vert\,\,\,\vert}}C-H} \;+\; \underset{\substack{\text{chlorine}}}{Cl-Cl} \;\longrightarrow\; \underset{\substack{\text{chloromethane}}}{H-\overset{\displaystyle H}{\underset{\displaystyle H}{\vert\,\,\,\vert}}C-Cl} \;+\; \underset{\substack{\text{hydrogen chloride}}}{H-Cl}
$$

When methane reacts with chlorine, a hydrogen atom
"changes places" with a chlorine atom. The covalent bond
between the carbon atom and one hydrogen atom in the
methane molecule is broken. A new covalent bond between the
carbon atom and a chlorine atom is formed. A chemical
reaction which involves molecular substances always involves
the breaking of covalent bonds and the formation of new ones.
We know that reactions between molecular compounds are slow
(Experiment 25(a)). This is because covalent bonds – which hold
the atoms in the molecule together – are strong.

The forces between molecules are weak. The forces within
molecules are strong.

The properties of water

We know that water has some curious properties. It is a
molecular substance whose molecules have a low molecular
mass, yet it is a liquid and not a gas. Compare propane (mass
compared with mass of one hydrogen atom = 30), which is a
gas, with water (mass compared with mass of one hydrogen
atom = 18), which is a liquid. Why is water a liquid when we
expect it to be a gas?

We also know that water is a particularly good solvent. It
dissolves ionic and many molecular compounds. Usually,
molecular solvents (such as ethanol) dissolve molecular solids
but do not dissolve ionic solids. How is water able to dissolve
both?

Molecular substances "like" other molecular substances, and
ionic substances "like" other ionic substances. We say that they
have an affinity for each other. So molecular solids dissolve in
molecular liquids. Water is therefore able to behave like a
molecular substance towards molecular solids. It also dissolves
ionic solids. So it is also able to behave like an ionic substance
towards ionic solids. Perhaps the reason behind this will also
explain the fact that water is a liquid.

A water molecule is composed of one oxygen atom and two
hydrogen atoms. Oxygen has six electrons in its outermost shell,
and a hydrogen atom has one. There is one covalent bond
between each hydrogen atom and the oxygen atom.

✕ electron originally from oxygen
● electron originally from hydrogen

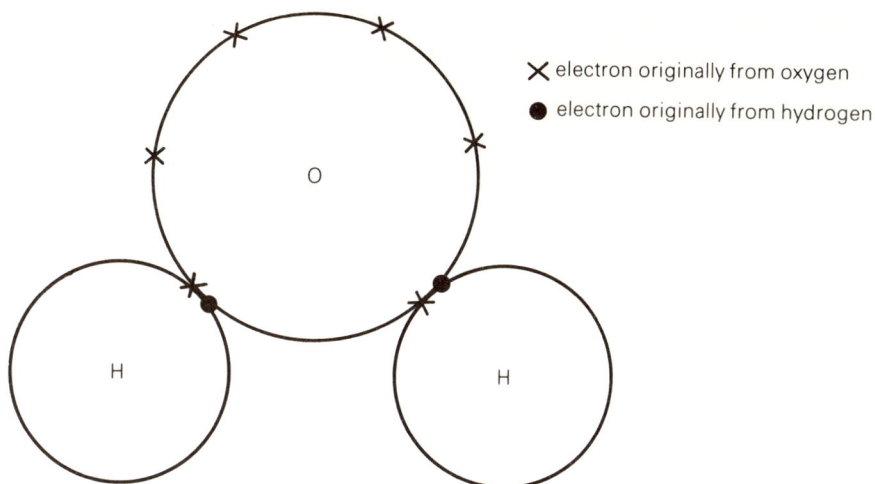

So the molecule looks like this:

O
/   \
H   H

It so happens that the oxygen atom attracts the electrons in the covalent bonds more than each hydrogen atom does. So the electrons which make up the bond are actually nearer to the oxygen atom than to the hydrogen atom. This means that the oxygen end of the molecule has more negative charge than expected and is therefore slightly negatively charged. The hydrogen end of the molecule has less negative charge than expected and is therefore slightly positively charged.

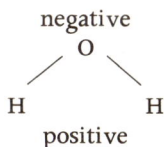

negative
O
/   \
H   H
positive

Although water molecules are molecular, then, they are also a little bit ionic. The positive end of one water molecule will attract the negative end of a neighbouring molecule, and so on. So there is a stronger force of attraction between molecules of water than between molecules such as methane which do not have oppositely charged ends. This extra attraction means that it is harder to separate molecules of water than would otherwise be the case. So water has a higher boiling point than expected. This makes water a liquid at room temperature and not a gas.

Because a water molecule is both molecular and ionic, it has an affinity for both molecular and ionic substances. So both ionic and molecular solids will dissolve in water.

Section E
Ionic Compounds

Topic 1
SALTS

E1.1 What are salts?

In Section C we learned that when a metal and a non-metal react together, an ionic compound is formed. We also learned that non-metals can react together forming molecular compounds.

In Section D we saw that some *elements* are molecular. This is because some non-metals can react with *themselves* forming molecules. But these molecules contain only one type of atom and so a compound has not been formed, and the substance is still an element.

Can we have ionic elements? A moment of thought will tell us that this is not possible. Non-metals can react with other non-metals, so a given non-metal can react with itself. But metals cannot react with other metals, they react with non-metals. *So a metal cannot react with itself.* It has to react with another element. Metals always form *compounds* when they react.

So *ionic substances are always compounds*, because they are never elements.

Most ionic compounds are *salts* – there are only a few which cannot be given this name. *A salt is an ionic compound which consists of a metal cation together with an anion which can be found in an acid.*

Anions which are found in acids consist of either a single non-metal or a non-metal combined with oxygen. For example, the following table gives the formulae of a number of anions found in common acids, with the non-metal underlined:

Acid	Formula	Anion
Hydrochloric	HCl	$\underline{Cl}^-$
Nitric	HNO_3	$\underline{N}O_3{}^-$
Sulphuric	H_2SO_4	$\underline{S}O_4{}^{2-}$
Phosphoric	H_3PO_4	$\underline{P}O_4{}^{3-}$

E1.2 Giving names to salts

Since all salts contain anions which can be found in acids, we can name a salt according to the name of its *parent* acid (the one which might have supplied the anion). The table on the next page gives the rules for going from the name of the acid to the name of the salt:

Name of acid		Name of salt	
Start	End	Start	End
Anything	-ic ⟶	Stays almost the same	-ate
Anything	-ous ⟶	Stays almost the same	-ite
Hydro-	-ic ⟶	Drops "hydro"	-ide

Examples

Sulphur*ic* acid ⟶ a sulph*ate*
Sulphur*ous* acid ⟶ a sulph*ite*
*Hydro*chlor*ic* acid ⟶ a chlor*ide*

If you know these rules you will be able to name almost all the salts you will come across in chemistry. You will find some questions at the end of the Topic which will give you practice in the rules and which will help you to remember them.

E1.3 How can we make salts?

We know that a salt must contain a metal cation and an anion (either a single non-metal or a non-metal combined with oxygen). So:

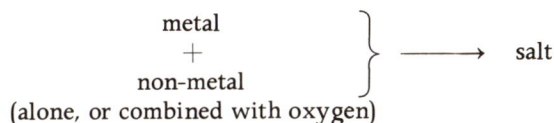

metal
+
non-metal
(alone, or combined with oxygen)   } ⟶ salt

Obviously, one way of making a salt is to react a metal directly with a non-metal. But are there other ways? We might try using *compounds* containing metals and non-metals. We know that acids must contain a non-metal since we know that:

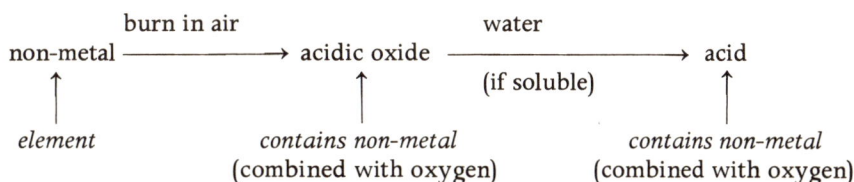

non-metal —— burn in air ⟶ acidic oxide —— water (if soluble) ⟶ acid
↑ element   ↑ contains non-metal (combined with oxygen)   ↑ contains non-metal (combined with oxygen)

For metals we know that:

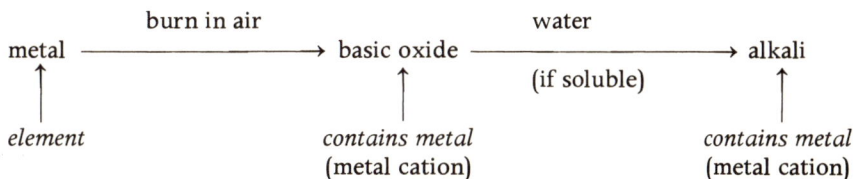

metal —— burn in air ⟶ basic oxide —— water (if soluble) ⟶ alkali
↑ element   ↑ contains metal (metal cation)   ↑ contains metal (metal cation)

So if we had to make a salt starting, for example, from a metal itself, we could try reacting it with an acidic oxide or with an acid as well as with a non-metal. You can now see that we have a large number of possible ways of making salts. For instance, it seems that an acid should react with an alkali to make a salt.

■ Write down all the different ways of making salts suggested by the information above. (There are nine ways altogether).

Not all of the nine ways work in practice, because it is necessary for water to be present for the reaction to work. The one exception to this is the reaction of a metal directly with a non-metal. Remembering that acids and alkalis both contain water, we can now decide which of the nine methods can actually be used to obtain salts:

Method	Works in practice?	Example
1 Metal + non-metal	✓	iron + chlorine ⟶ iron(III) chloride
2 Metal + acidic oxide	x (no water)	
3 Metal + acid	✓	magnesium + hydrochloric acid ⟶ magnesium chloride
4 Basic oxide + non-metal	x (no water)	
5 Basic oxide + acidic oxide	x (no water)	
6 Basic oxide + acid	✓	Copper(II) oxide + sulphuric acid ⟶ copper(II) sulphate
7 Alkali + non-metal	✓	Sodium hydroxide + chlorine ⟶ sodium chloride
8 Alkali + acidic oxide	✓	calcium hydroxide + carbon dioxide ⟶ calcium carbonate
9 Alkali + acid	✓	Potassium hydroxide + nitric acid ⟶ potassium nitrate

We are left with six ways of making salts. Three of these methods start from elements and three start from compounds.

E1.4 Making salts from elements

> You will see how experiments you have met before can be adapted in order to obtain the salts produced.

Making iron(III) chloride by direct combination

Each of the following reactions will give us a salt:

1 A metal + a non-metal
2 A metal + an acid
3 A non-metal + an alkali.

In fact, you have already met two of these reactions.

In Topic A3.3 we saw that the "halogens" will react with heated iron. When you saw chlorine react in this way, you may have noticed that a brown solid was left at the end. This was the salt iron(III) chloride. In order to collect the salt, we will obviously have to change the design of the apparatus.

When you saw this reaction before, you should have seen that once the iron and chlorine were reacting, dense brown smoke poured out of the test-tube. This too was iron(III) chloride – tiny particles of the solid being carried in the stream of chlorine. If we had heated the whole test-tube very strongly all the way through the experiment, we would have found no brown iron(III) chloride solid inside it at the end. All the salt would have been forced to pass out of the tube as "smoke".

We could then obtain the iron(III) chloride by allowing the particles to collect on the inside of a large glass vessel.

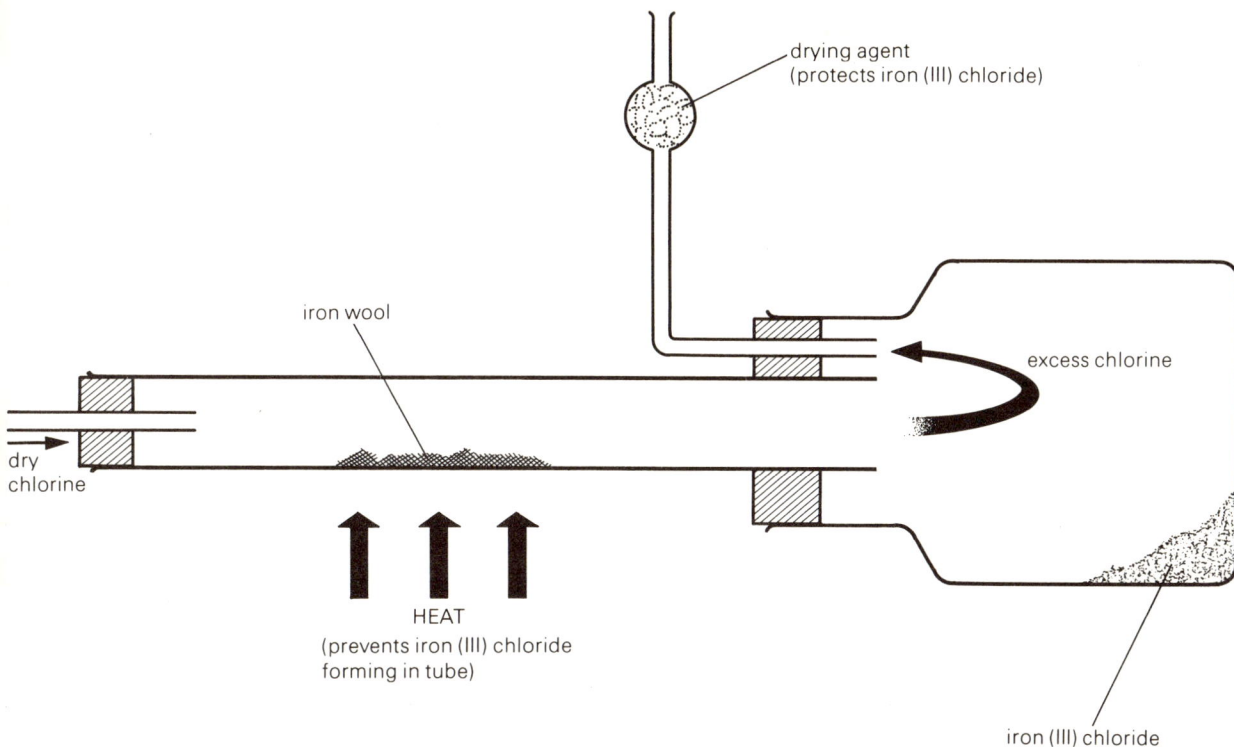

drying agent
(protects iron (III) chloride)

iron wool

dry
chlorine

excess chlorine

HEAT
(prevents iron (III) chloride
forming in tube)

iron (III) chloride

Iron(III) chloride reacts with the water-vapour in the air, so it has to be protected throughout the experiment, and the chlorine has to be dried before it is used. At the end of the experiment, the iron(III) chloride can be scraped off and placed in dry surroundings. The word-equation for this reaction is:

iron (s) + chlorine (g) ⟶ iron(III) chloride (s)

Because the elements iron and chlorine combine together directly, this method of preparation is called *direct combination*.

Making magnesium chloride from magnesium metal.

You have already come across the reaction between magnesium metal and hydrochloric acid in several places. You should remember that the word-equation is:

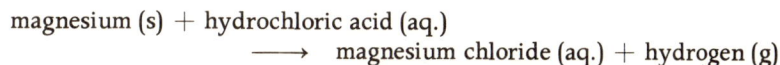

magnesium (s) + hydrochloric acid (aq.)
⟶ magnesium chloride (aq.) + hydrogen (g)

In Experiment 21 (Topic B4.2) you found out exactly how much magnesium reacted with a certain amount of hydrochloric acid. We can use this information to help us obtain the salt magnesium chloride.

Experiment 32

The following will be needed: measuring cylinder; 250 cm³ beaker; filter funnel; evaporating basin; filter paper; distilled water; magnesium ribbon (10 cm); approx. 0.5M hydrochloric acid (10 cm³).

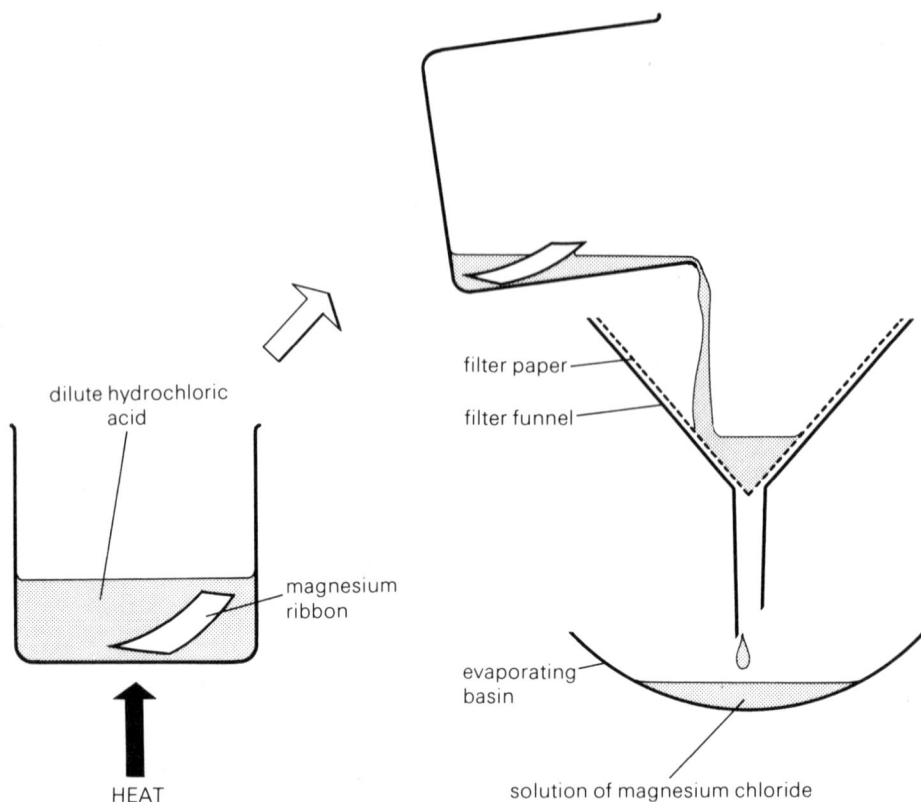

dilute hydrochloric acid

magnesium ribbon

HEAT

filter paper

filter funnel

evaporating basin

solution of magnesium chloride

Measure 10 cm³ of the hydrochloric acid provided into a 250 cm³ beaker. Clean the piece of magnesium ribbon, and cut it into 1 cm lengths.

Add five of these pieces of magnesium to the hydrochloric acid and wait for them to dissolve. If this takes a long time, heat gently.

Now add more pieces of magnesium, one at a time, heating the beaker all the time.

Do not allow the beaker to boil dry. If this is going to happen, add distilled water. (Keep the volume at about 10 cm³ all the time).

Continue until the last piece of magnesium that you have added does not dissolve. Filter the solution into a clean evaporating basin.

Heat the solution until solid appears around the edge of the solution, and then place the evaporating basin on one side to cool. (At this stage the experiment is best left for 2–3 days).

Filter off the crystals which have formed, and dry them carefully on a piece of filter paper. These crystals are pure magnesium chloride.

During and after the experiment, you should answer the following:
1   How did you know that five pieces of magnesium could be added straight away?
2   Why was it important that the experiment should be continued until some magnesium was left *unreacted* at the end?
3   What does the word *crystal* mean? Can you sketch a crystal of magnesium chloride and describe its shape in words?
4   Is magnesium chloride soluble in water?

The reaction between sodium hydroxide and chlorine

Although this reaction gives us sodium chloride, it is not a method which we would normally use to prepare it since we obtain other salts at the same time (from which the sodium chloride would have to be separated). In fact, sodium hydroxide reacts differently with chlorine, depending on whether it is hot or cold. The word-equations are:

cold
sodium hydroxide (aq.) + chlorine (g)
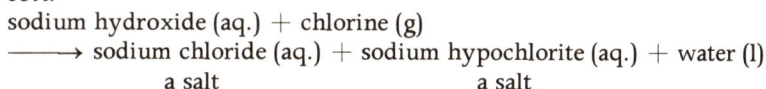
⟶ sodium chloride (aq.) + sodium hypochlorite (aq.) + water (l)
     a salt   a salt

hot
sodium hydroxide (aq.) + chlorine (g)
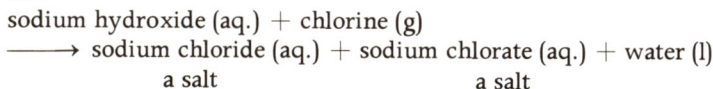
⟶ sodium chloride (aq.) + sodium chlorate (aq.) + water (l)
     a salt   a salt

Write word-equations for the reaction between potassium hydroxide and chlorine (a) in the cold and (b) when hot.

E1.5   Making salts from compounds

From our original list we are left with the following methods for preparing salts, each of which uses only compounds and not elements:

1   An acid + an alkali
2   An acid + a base
3   An acidic oxide + an alkali.

Of these, two are commonly used in the laboratory to prepare salts.

You will be given the opportunity of making salts by one or more methods as time allows.

Making sodium sulphate by neutralisation

You know that acids and alkalis "neutralise" each other. When they do so, a salt is always produced. To obtain the salt in pure form (that is, free from any acid or alkali), we have to react *exactly* the right amounts of acid and alkali together, so that none of either is left over at the end to contaminate the salt:

Too much
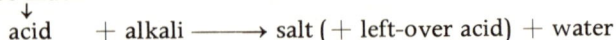
acid   + alkali ──────→ salt (+ left-over acid) + water

Too much
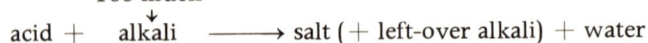
acid +   alkali   ──────→ salt (+ left-over alkali) + water

Exactly the right amounts
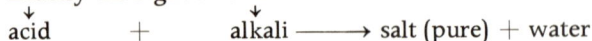
acid   +   alkali ──────→ salt (pure) + water

So we cannot just add acid and alkali in any amounts. We must find what the exact amounts are, and *then* react them together. We have, in fact, to do *two* experiments.

Experiment 33

The following will be required: burette; 250 cm³ beaker; measuring cylinder; evaporating basin; phenolphthalein solution; plenty of bench dilute sodium hydroxide solution and dilute sulphuric acid.

If you have not used a burette before, your teacher will show you how to handle this piece of apparatus before you start the experiment.

Using a measuring cylinder, put 25 cm³ of bench dilute sodium hydroxide solution into a beaker. To this add 2–3 drops of phenolphthalein indicator. This will give a pink colour.

Place bench dilute sulphuric acid in the burette in the way you were shown by your teacher. Note the reading on the burette. Carefully run the sulphuric acid into the sodium hydroxide solution. After each cm³ has been added, stop and swirl the contents of the beaker to make sure they mix.

Continue adding sulphuric acid *until the indicator just goes colourless*. Take the reading on the burette and calculate the volume of sulphuric acid in cm³ that was required to neutralise 25 cm³ of dilute sodium hydroxide. You are now left with a solution of sodium sulphate on its own, but contaminated with the indicator.

In order to obtain pure sodium sulphate repeat the whole experiment, but do not add phenolphthalein. Before you add the sulphuric acid, take the reading on the burette and calculate what the final reading will be when you have added the correct amount. You can then quickly run the acid down to this mark when you add it to the alkali.

Obtain crystals of sodium sulphate using an evaporating basin in the same way as you did for Experiment 32.

During and after the experiment you should try to answer the following:

stage 1

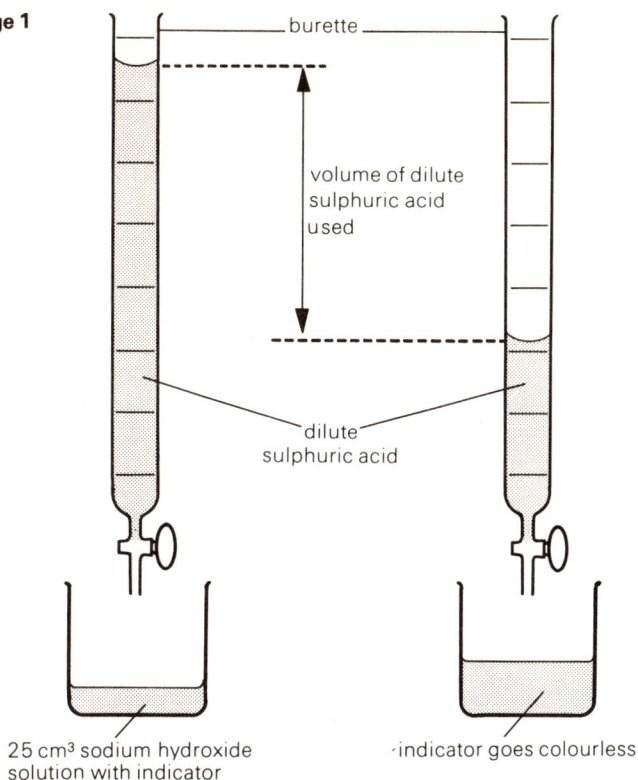

burette

volume of dilute
sulphuric acid
used

dilute
sulphuric acid

25 cm³ sodium hydroxide
solution with indicator

indicator goes colourless

stage 2

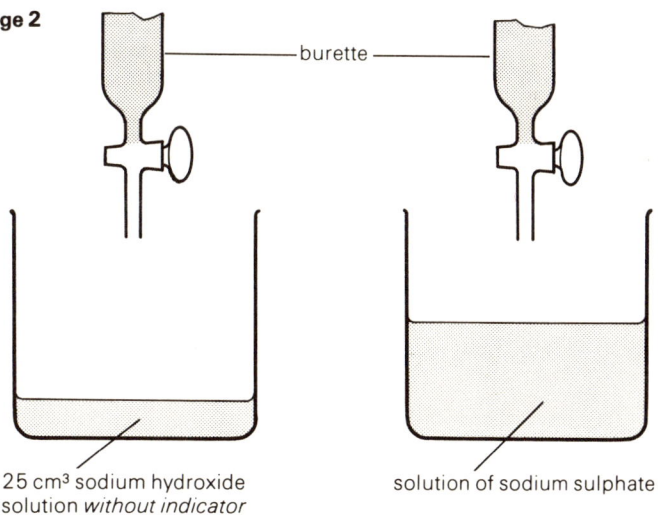

burette

25 cm³ sodium hydroxide
solution *without indicator*

solution of sodium sulphate

1   What colour is phenolphthalein indicator (a) in alkali and
(b) in acid?
2   Why could you stop adding the acid as soon as the indicator
turned colourless?
3   Sketch a crystal of sodium sulphate and try to describe its
shape in words.
4   Is sodium sulphate soluble in water?
5   Write a word-equation for the reaction.

Making copper(II) sulphate

An alkali is produced when a basic oxide dissolves in water. But some basic oxides are not soluble, and so we cannot make alkalis from them. However, they will themselves react with acids to give salts. The word-equation is always:

basic oxide + acid ⟶ salt + water

One such insoluble basic oxide is copper(II) oxide, and perhaps the most common salt made from it is copper(II) sulphate. When an insoluble basic oxide such as copper(II) oxide is used to make a salt, it is not possible to use an indicator to discover the exact point at which neutralisation occurs. So we must look for another way of making sure that the salt we make is pure.

One thing we could do would be to add the oxide to the acid until some was just left over at the end, showing that all the acid had reacted. If we then made crystals we would only have a tiny amount of oxide impurity, or we could even filter it off before making the crystals.

In practice it is a lot simpler if we carry out the whole reaction making sure that too much basic oxide is present, and give the acid present plenty of chance to react by heating. The oxide which is left over can easily be filtered off. So what we aim to do this time is:

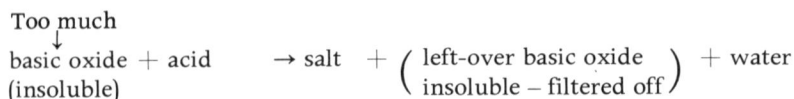

Too much
↓
basic oxide + acid   → salt + (left-over basic oxide) + water
(insoluble)   (insoluble – filtered off)

Experiment 34

The following will be needed: 250 cm³ beaker; evaporating basin; glass rod; measuring cylinder; copper(II) oxide powder; dilute sulphuric acid; filter funnel.

Place about 10 spatula-measures of copper(II) oxide powder in a beaker. Add 25 cm³ of dilute sulphuric acid and boil the mixture for 15 minutes. Throughout this time stir the mixture with a glass rod and make sure that some solid is always present. If necessary, add more copper(II) oxide powder.

At the end of 15 minutes, add about 10 cm³ of distilled water and filter off the unreacted copper(II) oxide, allowing the solution of copper(II) sulphate to run into an evaporating basin.

Obtain crystals and dry them as before.

25 cm³ dilute
sulphuric acid

0 spatula-
measures
of copper
(II) oxide

HEAT

remaining copper
(II)oxide

filter paper

filter funnel

solution of copper(II)sulphate

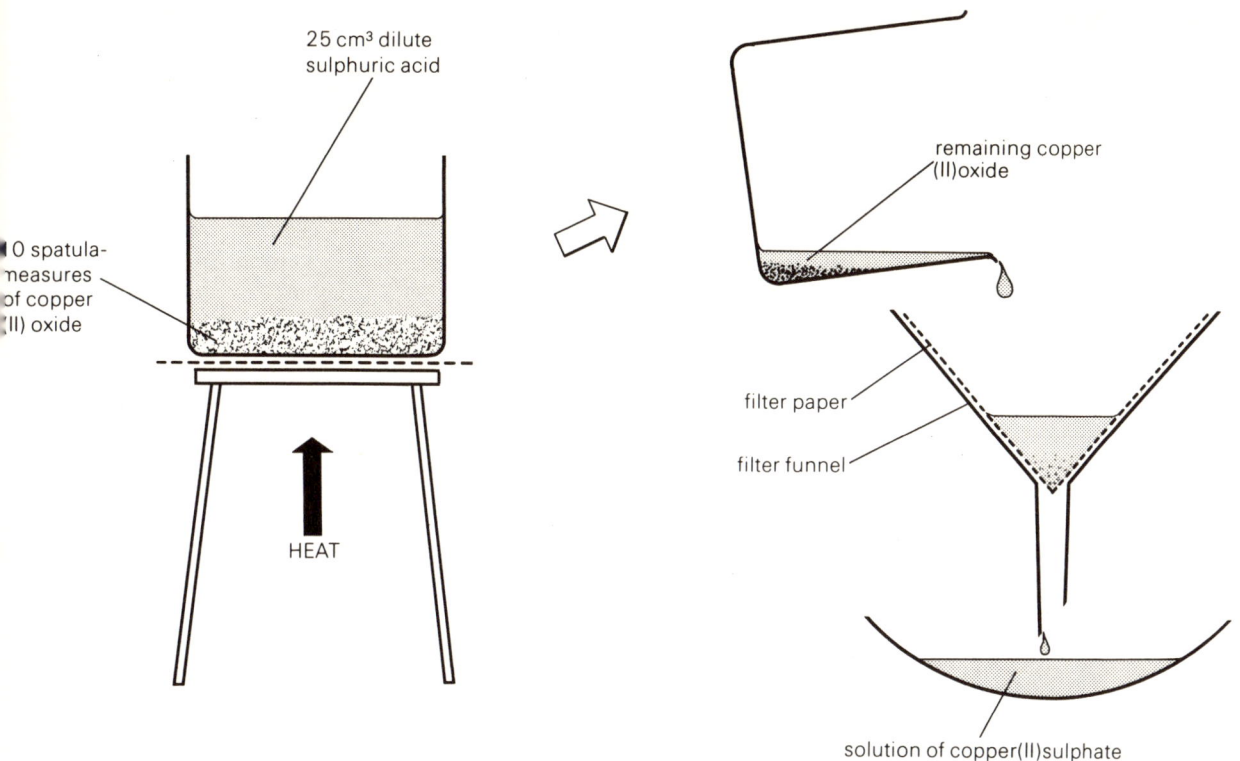

During and after the experiment you should try to answer the following:

1 Where do you think the blue colour of the copper sulphate comes from – the sulphuric acid or the copper(II) oxide?

2 Sketch a crystal of copper(II) sulphate and try to describe its shape in words.

3 Is copper(II) sulphate soluble?

4 Write a word-equation for the reaction.

The reaction between acidic oxides and alkalis

In general, this method is not used for preparing salts in the laboratory, since acidic oxides are usually gases which are soluble in water and form acids. It is easier to deal with acids than with the gases themselves, which often have a choking smell.

However, one good example of this type of reaction is the laboratory test for carbon dioxide. This reacts with calcium hydroxide forming the insoluble salt calcium carbonate, which is white. The word-equation is:

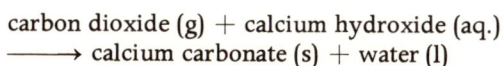

carbon dioxide (g) + calcium hydroxide (aq.)
$\longrightarrow$ calcium carbonate (s) + water (l)

As well as those already described, there are two further important ways of making salts from compounds.

Making salts from carbonates

Any carbonate will react with any acid in the following way:

carbonate + acid ⟶ salt + carbon dioxide + water

Since the only other product which could contaminate the salt is a gas, this reaction is very similar to the reaction between an insoluble basic oxide and an acid. All we have to do is to use a solid carbonate, and ensure that some is left at the end. You can try making copper(II) sulphate by this different route.

Experiment 35

The following will be needed: as for Experiment 34, replacing copper(II) oxide powder with copper(II) carbonate powder.

Carry out this preparation according to the instructions for Experiment 34, except for the following change: add the 25 cm³ of dilute sulphuric acid to the beaker, and *then* add to it the 10 spatula-measures of copper(II) carbonate powder, one spatula-measure at a time. Then stir and boil as before, always ensuring that some solid is present. Filter and obtain crystals.

Write a word-equation for the reaction.

Making salts by precipitation

All of the salts which have been made so far – except iron(III) chloride – have been soluble in water. Each has been made first as a solution in water and then obtained from this by crystallisation.

But what about salts which are insoluble (there are not very many of these, since we know that all salts are ionic and that most ionic compounds are soluble in water)? Obviously a solution of an insoluble salt cannot be made, so none of the methods used above would work. In fact, we have to "cheat" and make insoluble salts from other, soluble, ones.

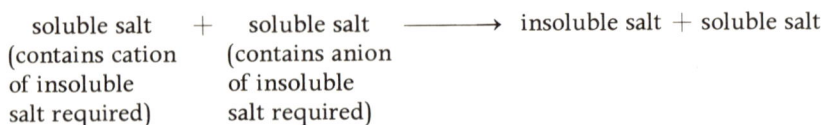

soluble salt   +   soluble salt   ⟶   insoluble salt + soluble salt
(contains cation   (contains anion
of insoluble   of insoluble
salt required)   salt required)

A typical example would be:

barium nitrate + sodium sulphate ⟶ barium sulphate + sodium nitrate
  (soluble)   (soluble)   (insoluble)   (soluble)

We can always find suitable soluble salts from which to make the insoluble salt we want, because *all nitrates are soluble and all sodium salts are soluble*. So if we start with the nitrate of the cation and react it with the sodium salt of the anion, we are bound to have two soluble starting materials and a soluble by-product (sodium nitrate) as well as the salt we want.

You saw four reactions in which insoluble salts were made in Topic C2.3, as a demonstration of the speed of reaction of ionic compounds. The solid which forms is called a *precipitate*, and so this method of making insoluble salts starting from soluble ones is called *precipitation*. All insoluble salts have to be made in this way.

You will find a set of rules for remembering which salts are insoluble in the Topic Summary.

You can now carry out the preparation of an insoluble salt, barium sulphate, yourself.

Experiment 36

The following will be needed: two 250 cm³ beakers; filter-funnel; filter-paper; plenty of distilled water; test-tubes and test-tube rack; two teat-pipettes; 0.2M barium nitrate solution (20 cm³); 0,2M sodium sulphate solution 20 cm³).

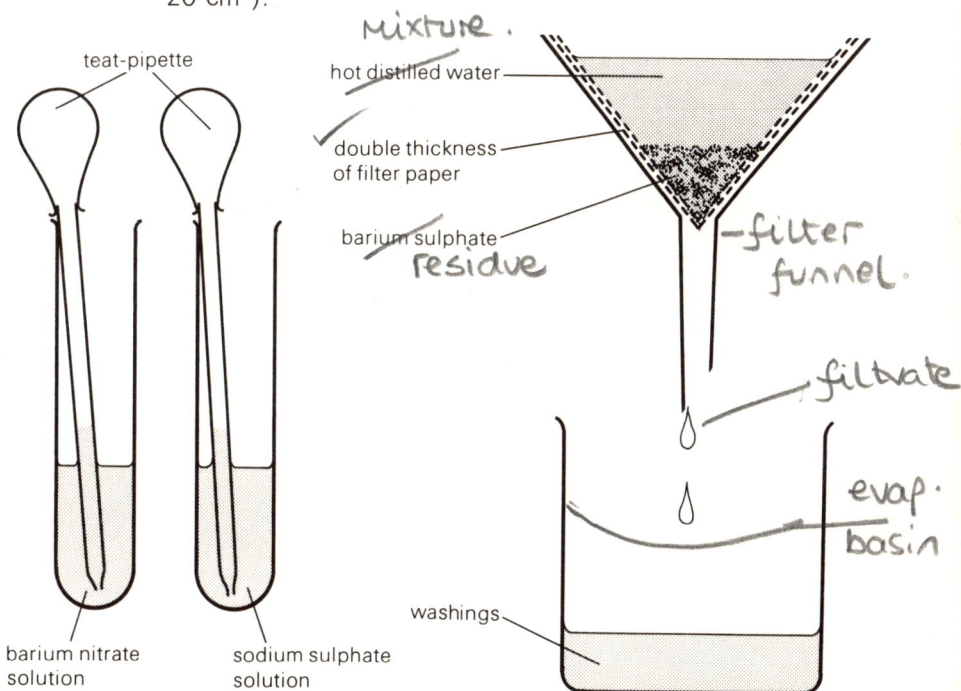

teat-pipette

Mixture .

hot distilled water

double thickness of filter paper

barium sulphate residue

filter funnel.

filtrate

evap. basin

washings

barium nitrate solution

sodium sulphate solution

Take the two starting materials (solutions) back to your bench and place a little of each (a depth of 2–3 cm) in two separate test-tubes. Into each of these put a *clean* teat-pipette, and mark each solution so that you remember which is which.

Now add together the remainder of the two solutions in the beakers, making insoluble barium sulphate. Since the starting materials and the by-product are soluble, you can simply filter off the barium sulphate. You should use a double thickness of filter-paper as barium sulphate is a very fine powder.

The precipitate will now be contaminated with traces of each of the three soluble materials present. These can be removed by washing with *hot* distilled water. Allow one portion of hot distilled water to drain through the filter-paper. Add more, and this time collect a few drops of the washings as they come out of the filter paper in each of two clean test-tubes. To one of these add two drops of barium nitrate solution using the teat-pipette. *If a white precipitate forms, this shows that the washings contain sodium sulphate and that the precipitate is not yet clean.* Return the teat-pipette to the correct tube.

To the other sample of the washings add two drops of sodium sulphate solution using the teat-pipette. *If a white precipitate forms, this shows that the washings contain barium nitrate and that the precipitate is not yet clean.* Return the teat-pipette to the correct tube.

Continue to add distilled water, collect some of the washings in *clean* test-tubes and carry out the two tests, until neither gives a white precipitate or cloudiness. Only at this point is the precipitate clean.

Transfer the clean precipitate to a dry piece of filter-paper and dry it by pressing a further piece of dry filter-paper onto its surface with the back of a spatula. Examine the barium sulphate solid.

During and after the experiment you should try to answer the following:

1   Why was the distilled water heated before it was used to wash the precipitate?

2   What was the white "cloudiness" produced in the tests before the precipitate was clean?

3   Why did *both* tests have to be carried out on the washings – wouldn't one of these on its own have been just as good?

4   Describe the barium sulphate you produced. Is it "crystalline" (in the form of crystals)?

Topic summary and learning

1   A salt is an ionic compound which consists of a metal cation together with an anion which can be found in an acid.

2   We can name salts according to the name of their "parent" acid, using the following rules (this is one method of naming – you will learn an even more systematic method later in your study of chemistry):

Name of acid		Name of salt	
Start	End	Start	End
Anything	-ic ⟶	Stays almost the same	-ate
Anything	-ous ⟶	Stays almost the same	-ite
Hydro-	-ic ⟶	Drops "hydro"	-ide

3 There are a number of ways of making salts. We can start either from elements or from compounds. The following types of reactions are the most common methods:

From elements

a) metal + non-metal ⟶ salt only (*direct combination*)
b) metal (above lead in the reactivity series) + acid ⟶ salt + hydrogen

From compounds

c) acid + alkali ⟶ salt + water   (*neutralisation*)
d) acid + insoluble base ⟶ salt + water   (*neutralisation*)
e) acidic oxide + alkali ⟶ salt + water   (*neutralisation*)
f) acid + carbonate ⟶ salt + carbon dioxide + water   (*neutralisation*)
g) soluble salt + soluble salt ⟶ insoluble salt + soluble salt

                                               (*precipitation*)

4 Soluble salts are made by a method which involves crystallisation. Insoluble salts are made by precipitation.

soluble salts	*insoluble salts*
All sodium, potassium and ammonium salts	
All nitrates	
All chlorides except . . .	silver chloride, lead chloride
All sulphates except . . .	barium sulphate, lead sulphate
	All carbonates except those of sodium, potassium and ammonium.

5 Most salts are soluble and form crystals. Crystals have a regular shape, with straight edges and flat surfaces.

Learn the information contained in the Topic Summary.
You should also remember the way in which each of the salts prepared during this Topic was obtained in a pure state.
You should be able to write a word-equation for each of the reactions carried out during the Topic.

Questions

1 Give a name to each of the following salts:
a) the sodium salt of carbonic acid
b) the calcium salt of sulphuric acid
c) the potassium salt of nitrous acid
d) the magnesium salt of phosphoric acid
e) the sodium salt of hydrobromic acid.

2 Give the name of the acid from which each of these salts could be obtained:
a) lithium nitrate
b) zinc chloride
c) lead iodide
d) sodium sulphate
e) sodium sulphite.

3   Complete each of the following word-equations:
a)   magnesium (s) + sulphuric acid (aq.) →
b)   zinc (s) + nitric acid (aq.) →
c)   lead oxide (s) + nitric acid (aq.) →
d)   potassium hydroxide (aq.) + hydrochloric acid (aq.) →
e)   calcium carbonate (s) + sulphuric acid (aq.) →
f)   sodium chloride (aq.) + silver nitrate (aq.) →

4   Write a word equation to represent a suitable reaction for making each of the following salts:
a)   copper(II) nitrate
b)   potassium nitrate
c)   lead chloride
d)   calcium carbonate
e)   sodium chloride
f)   zinc chloride
g)   sodium sulphite
h)   calcium chloride
i)   magnesium nitrate
j)   lead sulphate.

Topic 2
PASSING ELECTRICITY THROUGH SALTS

E2.1 Electrolysis of molten salts

You will electrolyse molten lead bromide.

In Experiment 23 (Topic C1.2) you found that ionic compounds would conduct electricity when melted (molten). Since salts are ionic, they will all conduct electricity when molten.

You can now repeat the electrolysis of molten lead bromide, but this time pay more attention to the changes which take place as the current passes through the molten salt.

Experiment 37

The following will be needed: power pack or batteries (6 volts); connecting wires; crocodile clips; carbon rods; small crucible; pipe-clay triangle; lead bromide.

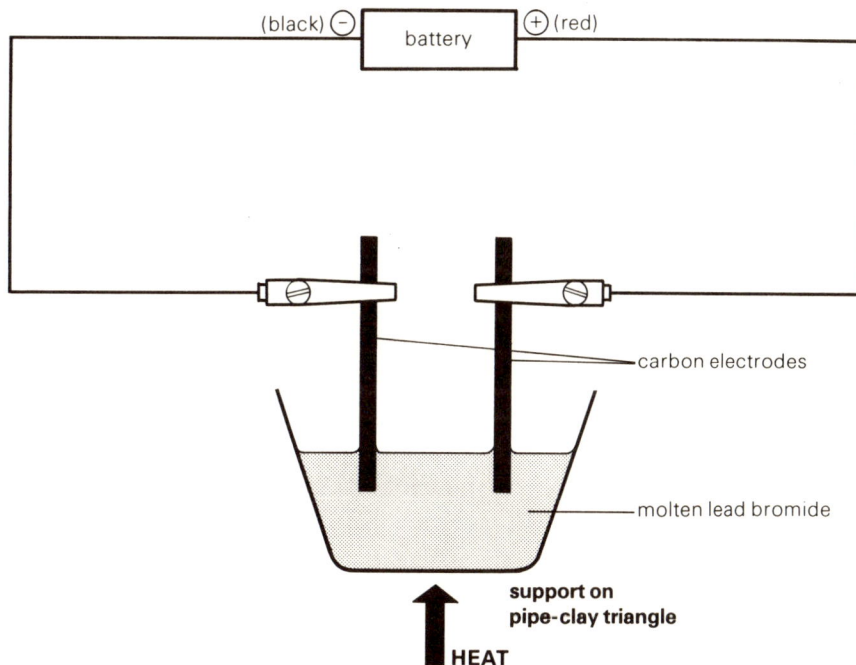

Your teacher will show you how to use the pipe-clay triangle with the small crucible.

Half fill the crucible with solid lead bromide, and melt it using the heat from a Bunsen burner.

Set up the electrical circuit to pass current through the molten lead bromide using carbon electrodes. (*Note:* no bulb is included in the circuit, since we already know that molten lead bromide conducts electricity.) *Be careful to keep the electrical leads away from the heat of the Bunsen. The lead bromide must be heated throughout the experiment in order to keep it molten.*

When you are ready, pass current through the molten salt, and continue this for 10 minutes. Observe and make a note of any visible changes which occur.

At the end of 10 minutes, remove the electrodes and allow them to cool. Break open the solid lead bromide in the area around the negative electrode.

During the electrolysis, you should have seen brown bromine vapour produced from the positive electrode. Under the negative electrode you should have found a bead of lead. *When a molten salt is electrolysed, it is split into a metal and a non-metal again. The metal is released at the negative electrode and the non-metal is released at the positive electrode.*

E2.2 Electrolysis of salts in solution

In Topic A2.3, electricity was passed through some salts in solution. It was seen that, as for molten salts, it is *possible* to get the metal of the salt at the negative electrode, and the non-metal of the salt at the positive electrode. However, the examples used in Topic A2.3 were rather special cases. In fact it is very rare for the substances produced when a *solution* of a salt is electrolysed to be identical with those which would be produced if the *molten* salt is electrolysed.

The reason for this is very simple. In a solution of a salt we also have water, and water can be split up by electrolysis into hydrogen (at the negative electrode) and oxygen (at the positive electrode).

There are some simple rules to decide exactly what happens for any salt in solution.

What happens at the negative electrode. At the negative electrode, the element which is lowest in the metal reactivity series is released.

Metal reactivity series

Group 1 metals and magnesium
Carbon
Zinc
Iron
Lead
Hydrogen
Copper
Silver

So hydrogen is always released, unless a copper or a silver salt in solution is electrolysed. For example, if a solution of sodium chloride is electrolysed, hydrogen is released at the negative electrode. But if a solution of copper(II) sulphate is electrolysed, copper is released at the negative electrode.

What happens at the positive electrode. At the positive electrode, oxygen is always released unless the solution is a concentrated solution of a chloride. In this case alone, chlorine gas is released.

You can now electrolyse some salt solutions to see whether these rules hold true.

Experiment 38

The following will be needed: power pack or batteries (12 volts); connecting wires; container with two fixed carbon electrodes; bungs for test-tubes; wooden spills; litmus paper; very concentrated solution of common salt; copper(II) sulphate solution; sodium sulphate solution; magnesium nitrate solution.

In each case, use the apparatus described in Experiment 5 (Topic A2.3), collecting and testing any gases produced.

Before you start to pass electricity, write down the elements which the rules above predict will be released at each electrode for the salt solution under test. Decide whether the rules hold true as a result of what you have seen yourself.

If copper is deposited on an electrode, clean it off at the end of your experiment.

No matter how many salt solutions are tested, the rules above are always obeyed.

Topic summary and learning

1   When a molten salt is electrolysed, it is split into a metal and a non-metal again. The metal is released at the negative electrode and the non-metal is released at the positive electrode.

2   When a salt in solution is electrolysed, hydrogen is always released at the negative electrode unless a copper or silver salt is being electrolysed. In this case, the copper or silver is itself released.

At the positive electrode, oxygen is always released unless the solution is a concentrated solution of a chloride. In this case alone chlorine gas is released.

Learn the information contained in the Topic Summary. You should also learn the detailed rules for which element is released at the negative electrode when the solution of a salt is electrolysed.

Questions

Predict which elements will be produced at the two electrodes when the following are electrolysed:

1   molten lead chloride
2   a solution of lead nitrate
3   a concentrated solution of potassium chloride
4   a dilute solution of copper(II) chloride
5   a concentrated solution of copper(II) chloride
6   molten silver chloride
7   a mixture of zinc nitrate and silver nitrate in solution
8   a mixture of magnesium sulphate and zinc sulphate in solution
9   a solution of silver nitrate
10   a solution of iron(II) nitrate.

Solid salts do not conduct electricity. Molten salts and salts in solution do. When a molten salt conducts electricity a chemical change takes place – a metal is produced at the negative electrode and a non-metal is produced at the positive electrode.

The salt must be in liquid form so that the ions which make it up are free to move. The ions in a solid salt cannot move, and that is why it cannot conduct electricity.

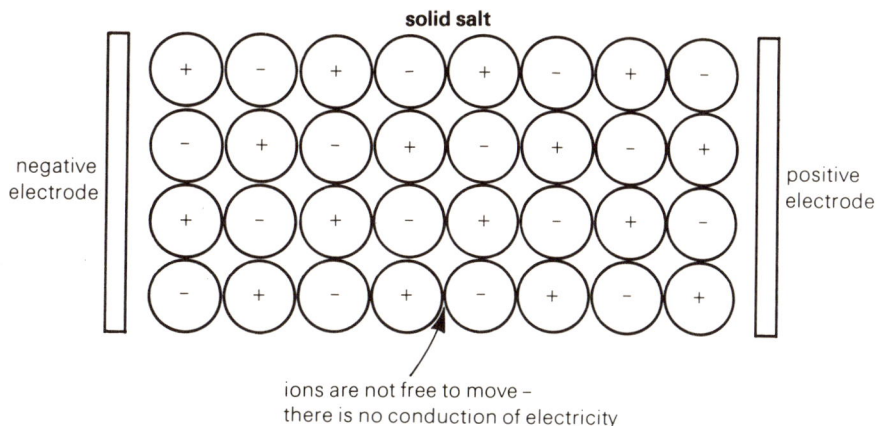

solid salt

negative electrode

positive electrode

ions are not free to move –
there is no conduction of electricity

The electric current is passed through the electrolyte by the movement of ions, and by the chemical changes that take place.

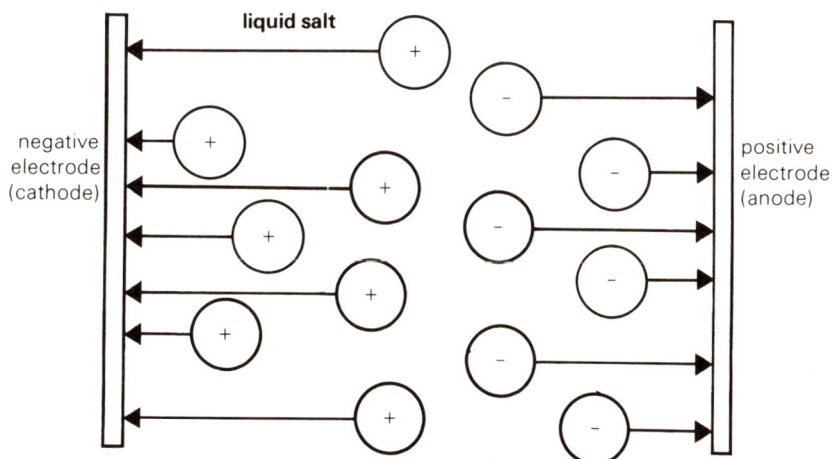

liquid salt

negative electrode (cathode)

positive electrode (anode)

The cations (positively charged ions) are attracted by the negatively charged electrode, and move towards it. For this reason the negatively charged electrode is called the cathode. The positively charged electrode is called the anode because the anions (negatively charged ions) move towards it. (To avoid confusion, remember: cations are pussitive.)

So, negative charge moves in one direction, and positive charge moves in the opposite direction. But what happens when the ions reach the electrodes? Let us take the molten salt lead bromide as an example. When molten lead bromide is electrolysed, the lead cations move towards the cathode and the bromide anions move towards the anode.

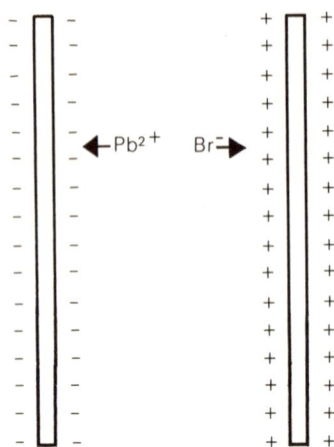

At the cathode, the lead ions (which have two positive charges) meet negatively charged electrons. Two electrons neutralise the two positive charges:

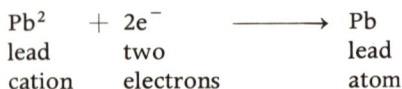

$$Pb^2 \; + \; 2e^- \longrightarrow Pb$$

| lead | two | lead |
| cation | electrons | atom |

The lead cation is discharged, and a lead atom is produced. So the element lead is produced at the cathode.

At the anode, the bromide ion (which has one negative charge) gives up an electron to the positively charged electrode:

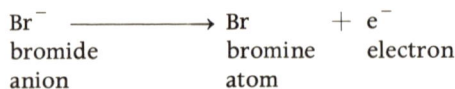

$$Br^- \longrightarrow Br \; + \; e^-$$

| bromide | bromine | electron |
| anion | atom | |

But we know that the element bromine is diatomic. Two bromine atoms join up producing a bromine molecule:

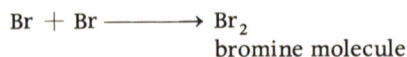

$$Br + Br \longrightarrow Br_2$$

bromine molecule

Overall, the process at the anode is:

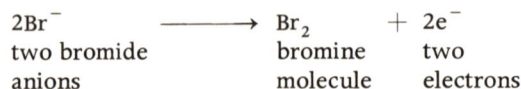

$$2Br^- \longrightarrow Br_2 \; + \; 2e^-$$

| two bromide | bromine | two |
| anions | molecule | electrons |

The bromide anions also lose their charge – they are discharged, and a bromine molecule is produced. So the element bromine is produced at the anode.

If we look at the process overall we can see that:

at the cathode	*at the anode*
lead liberated	bromine liberated
electrons used up	electrons produced.

So it seems as though the electrons go into the electrolyte at the cathode, and come out of it at the anode. The electrolyte conducts electricity, and is changed chemically into its elements as it does so.

When salts in solution are electrolysed, the situation is a little more complicated. Instead of there being just one anion and one cation present, there are two of each. The second anion and cation are provided by the water which is also now present. Water is always split up into ions to a small extent:

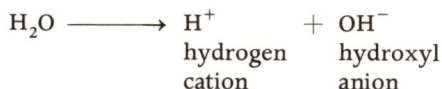

$$H_2O \longrightarrow H^+ \quad + \quad OH^-$$

hydrogen	hydroxyl
cation	anion

So suppose we pass an electric current through a solution of sodium chloride. There are two cations present (Na^+ from sodium chloride and H^+ from water). We know that it is the H^+ cation which is discharged (giving the element hydrogen), and that this is because hydrogen is lower in the metal reactivity series than sodium. But why should this be?

The metal reactivity series tells us how readily the metallic element reacts. We also know that when a metallic element (a metal atom) reacts, it forms an ionic compound (containing a metal cation). How easily the metal reacts depends on how easily the change

metal atom $\longrightarrow$ metal cation

takes place. The more easily it takes place, the more reactive the metal is, and the higher it is in the reactivity series.

In electrolysis, we are carrying out the reverse process. The ease of electrolysis is therefore the ease of the process:

metal cation $\longrightarrow$ metal atom.

The lower an element is in the reactivity series the easier this process is. So the cation which is lower in the reactivity series is the one which is discharged if there is a choice of two.

```
                        ↑ Alkali metals │
                        │ Magnesium     │
                        │ Zinc          │
atom ——→ cation         │ Iron          │  cation ——→ atom
         easier         │ Lead          │           easier
                        │ Hydrogen      │
                        │ Copper        │
                        │ Silver        ↓
```

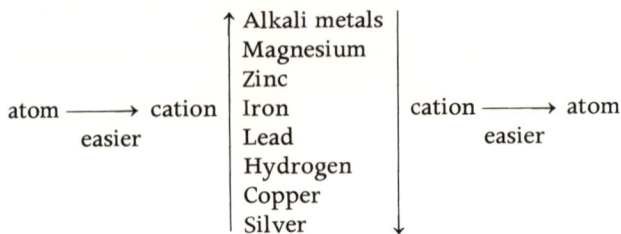

Because hydrogen is lower in the series than all metals except copper and silver, it is discharged in preference to these other metals. Copper and silver are the only metals which can be liberated at cathodes by the electrolysis of salts in solution.

When a salt in solution is electrolysed, the hydroxyl ion moves to the anode. The hydroxyl ion is discharged giving the element oxygen. The only time another anion is discharged in preference to the hydroxyl anion is when there is a very concentrated solution of a chloride. In this case, chlorine gas is liberated at the anode. In all other cases of electrolysis of a salt in solution, oxygen is liberated at the anode.

REQUIREMENTS FOR EXPERIMENTS

Quantities indicated are those needed by each pupil or group of pupils. D indicates that the experiment is demonstrated by the teacher.

Experiment

1 D — Samples of the elements sodium, potassium, magnesium, aluminium, iron, zinc, mercury, carbon, sulphur, bromine and iodine; gas jars containing chlorine and oxygen.

2 — Power pack or batteries (6 volts); connecting wires; crocodile clips; probes; bulb and bulb-holder; samples of magnesium, iron, copper, aluminium, graphite and sulphur (lumps).

3 — Length of magnesium ribbon (2–3 cm); test-tube containing carbon dioxide; small piece of calcium; powdered sulphur; tongs; emery paper; spatula; broken porcelain; 4 pieces of Universal Indicator paper; small beaker; safety spectacles.

4 — Power pack or batteries (12 volts); connecting wires; crocodile clips; old graphite electrodes; small beaker; copper(II) sulphate solution; emery paper.

5 — Power pack or batteries (12 volts); connecting wires; container with two fixed graphite electrodes; very strong sodium chloride solution marked "concentrated brine"; test-tubes; Litmus paper; wooden splint; bungs for test-tubes.

6 — Charcoal block; blow pipe; lead oxide (litharge); safety spectacles.

7 D — Silver nitrate solution; small strip of magnesium ribbon; glass petri dish; overhead projector (if available).

8 D — Samples of lithium, sodium and potassium; 3 gas-jars containing chlorine; 3 deflagrating spoons; 2 beakers containing small quantities of water; Universal Indicator paper.

9 — 2 small pieces of lithium; broken porcelain; tongs; beakers; Universal Indicator paper; safety spectacles.

10 D — Chlorine generator; bromine; iron wool; test-tube with hole at the end; 2 boiling tubes; test-tube and bung; Universal Indicator solution.

Experiment

11   2 crystals of iodine; iron wool; test-tube and bung; boiling tube; Universal Indicator paper.

12 D   Powdered zinc; dilute sulphuric acid; ammonium metavanadate; 100 cm³ measuring cylinder.

13 D   Weighed samples of iron filings (14 g) and sulphur powder (8 g); mortar and pestle; hard glass test-tubes.

14   Clean magnesium ribbon in 1 cm lengths; boiling tube; 20 cm³ measuring cylinder; stirring thermometer; stop clock (if available); 100 cm³ beaker.

15   150 cm³ 0.2M sodium thiosulphate solution; 100 cm³ beaker; stopclock (if available); dilute hydrochloric acid; 2 100 cm³ measuring cylinders.

16(a)   1 cm length of clean magnesium ribbon; dilute hydrochloric acid; 100 cm³ beaker.

16(b)   10 cm length of clean magnesium ribbon; magnesium powder; 100 cm³ beaker; access to top-pan balance.

17   Solutions of potassium chloride, potassium bromide and potassium iodide in water; chlorine water; bromine water; 6 test-tubes; test-tube rack; teat-pipette.

18   Samples of the following metals: magnesium (ribbon), lead (foil), iron (filings), zinc (foil), copper (foil). The following aqueous solutions: magnesium chloride, iron(II) sulphate, zinc sulphate, copper(II) sulphate, silver nitrate. Microscope slides or a white tile; a glass rod or teat-pipette.

19   Dilute hydrochloric acid; short piece of magnesium ribbon; zinc foil; a small piece of iron; copper foil.

20 D   Sodium chloride solution; silver nitrate solution; 250 cm³ conical flask with tight-fitting bung; small test-tube.

21   10 cm length of clean magnesium ribbon; 15 cm³ approximately 0.5M hydrochloric acid; 20 cm³ measuring cylinder; Universal Indicator paper.

22   Power pack or batteries (12 volts); connecting wires; crocodile clips; container with two fixed graphite electrodes; water acidified with dilute sulphuric acid or dilute hydrochloric acid, marked "acidified water."

Experiment

23 Power pack or batteries (6 volts); connecting wires; crocodile clips; carbon rods; bulb and bulb-holder; The following solids: lead bromide, potassium iodide, paraffin wax. Ethanol.

24 Power pack or batteries (6 volts); connecting wires; crocodile clips; carbon rods; bulb and bulb-holder. The following compounds: sodium chloride, potassium nitrate, ethanol, glucose, potassium iodide, paraffin wax, ethyl acetate, petroleum ether. Distilled water.

25(a) D "Quickfit" pear-shaped flask and reflux condenser; anti-bumping granules; test-tube rack; concentrated sodium hydroxide solution (10 cm³); potassium permanganate solution; benzyl alcohol (phenylmethanol); concentrated hydrochloric acid.

25(b) D The following aqueous solutions: sodium chloride, barium nitrate, lead nitrate, silver nitrate, sodium sulphate, potassium iodide, potassium chromate. Test-tube rack.

26 D 2 gas-jars containing carbon dioxide; magnesium ribbon; tongs.

27 The following solids: sodium carbonate, copper(II) carbonate, sodium chloride, copper(II) chloride, sodium sulphate, copper(II) sulphate. Dilute hydrochloric acid; limewater.

28 D Filter-pump; Dreschel bottle (or similar); thistle-funnel; candle; watch-glass; sucrose.

29 The following aqueous solutions: calcium hydrogencarbonate, calcium nitrate, calcium sulphate, sodium hydrogencarbonate, sodium nitrate, sodium sulphate. Distilled water; soap solution.

30 Sample of crude oil; test-tube with side-arm; thermometer (0°–360°C) in bung for test-tube; delivery tube and connector; Rocksill wool.

31 5% solutions of hexamethylene diamine (in water) and adipyl chloride (in tetrachloromethane); 10 cm³ beaker (or crucible), tongs; glass rod.

32 10 cm length magnesium ribbon; 10 cm³ approximately 0.5M hydrochloric acid; 20 cm³ measuring cylinder; 250 cm³ beaker; filter funnel; filter paper; evaporating basin; distilled water.

33 Burette; 250 cm³ beaker; 100 cm³ measuring cylinder; evaporating basin; phenolphthalein solution; adequate supply of bench dilute sodium hydroxide solution and dilute sulphuric acid.

Experiment

34 Copper(II) oxide powder; dilute sulphuric acid; 250 cm³ beaker; evaporating basin; 100 cm³ measuring cylinder; glass rod; filter paper; filter funnel.

35 Copper(II) carbonate powder; dilute sulphuric acid; 250 cm³ beaker; evaporating basin; 100 cm³ measuring cylinder; glass rod.

36 20 cm³ 0.2M barium nitrate solution; 20 cm³ 0.2M sodium sulphate solution; 2 250 cm³ beakers; filter funnel; filter paper; 2 teat-pipettes; test-tube rack; distilled water.

37 Power pack or batteries (6 volts); connecting wires; crocodile clips; carbon rods; small crucible; pipe-clay triangle; lead bromide.

38 Power pack or batteries (12 volts); connecting wires; crocodile clips; container with two fixed graphite electrodes; bungs for test-tubes; wooden spills; litmus paper; concentrated brine; copper(II) sulphate solution; sodium sulphate solution; magnesium nitrate solution.

GLOSSARY OF TERMS AND TOPIC INDEX

Topic

A1.3	*Acid* – an aqueous solution whose pH is less than 7
A1.3	*Alkali* – an aqueous solution whose pH is greater than 7
C1.2	*Anion* – an ion which has a negative charge
	Anode – the positively charged electrode in electrolysis
C1.2, D1.2	*Atom* – the smallest particle of an element
A1.3, E1.3	*Basic oxide* – an oxide which reacts with an acid to form a salt
	Boiling – the process of changing a liquid into a gas at its boiling point·
C2.2	*Boiling point* – the temperature at which boiling takes place for a given substance
	Cathode – the negatively charged electrode in electrolysis
C1.2	*Cation* – an ion which has a positive charge
	Chemical property – a chemical reaction typical of a chemical substance
	Chemical reaction – a change which results in the formation of new chemical substances
A1.1, C1.1	*Compound* – a pure substance made of two or more elements combined together chemically
B2.3	*Concentration* – the amount of a substance in a given volume of a solution of it
	Condensation – the process of changing a gas into a liquid below its boiling point
C2.2	*Density* – the amount of mass in a given volume of a substance
	Distillation – boiling followed by condensation
A1.3	*Electrical conductivity* – the ability to conduct electricity
A2.3, E2.1	*Electrode* – the electrical contact between an electrolyte and the battery in electrolysis
A2.3, E2.1	*Electrolysis* – the passage of electricity through a chemical substance, when this is accompanied by a chemical change
E2.1	*Electrolyte* – a chemical substance which conducts electricity which is chemically changed when it does so
A1.1	*Element* – a chemical which cannot be split into simpler substances using a chemical reaction
B1.1	*Endothermic process* – one which takes heat in from the surroundings
B1.1	*Exothermic process* – one which gives heat out to the surroundings
D3.2	*Fractional distillation* – the process of separating a complicated mixture of substances into fractions (or, in a simple case, into individual compounds) by distillation
E2.1	*Fused* – melted (molten)
A1.2	*Gas* – a substance in the gaseous state above its boiling point

ANSWERS TO QUESTIONS

Topic A1
1 –   2 –   3 Nitrogen: non-metal   Lithium: metal
Phosphorus: non-metal   Calcium: metal
4 –

Topic A2
1 Sodium, magnesium, aluminium   2 Chlorine, in each case
3 A   4 B   5 B   6 C   7 D   8 –

Topic A3
1 (a) More reactive   (b) Yes – very easily   (c) Very easily
in air. Very easily in chlorine   (d) It would move about on the
surface. The gas produced would catch fire. An alkaline solution
would be produced.
2 (a) More reactive   (b) Yes – producing an acid   (c) Yes –
very easily
3 E   4 Gold   5 (a) Copper   (b) Iron   (c) Gold
(d) Iron

Topic B1
1 (a) E   (b) B and D   (c) A and D   2 B
3 (a) Sulphur (s) + oxygen (g) → sulphur dioxide (g)
   (b) Sodium oxide (s) + water (l) → sodium hydroxide (aq.)
   (c) Sulphur dioxide (g) + water (l) → sulphurous acid (aq.)
   (d) Carbon (s) + lead oxide (s) → lead (s) + carbon dioxide (g)
   (e) Silver nitrate (aq.) + magnesium (s)
              → silver (s) + magnesium nitrate (aq.)
   (f) Carbon (s) + iron oxide (s) → iron (s) + carbon dioxide (g)
   (g) Lithium (s) + chlorine (g) → lithium chloride (s)
   (h) Sodium (s) + water (l)
              → hydrogen (g) + sodium hydroxide (aq.)
   (i) Iron (s) + chlorine (g) → iron(III) chloride (s)
   (j) Iron (s) + sulphur (s) → iron(II) sulphide (s)

Topic B2
1 C   2 –   3 –   4 Less rapidly   5 D   6 C £6000

Topic B3
1 –   2 Zinc, iron, lead, copper, silver   3 Copper, silver
4 A, B and E   5 B and C

Topic B4
1 –   2 (a) 1 g   (b) 0.1 g   (c) 12.3 g   3 (a) 32 g
(b) 8 g   (c) 40 g
4 30 cm³   5 20 cm³   6 –

Topic C1
1 –

Topic C2
1 Ionic. High M.P., B.P. and density. Conducts electricity when molten and when in solution. Rapid reactions.
2 Molecular. Low M.P. and B.P. Low solubility in water. Does not conduct electricity when molten or when in solution.
3 It is ionic.   4 –   5 Copper chloride: ionic   Methyl acetate: molecular   Isopentane: molecular   Calcium phosphate: ionic
6 (a)   Sodium chloride (aq.) + silver nitrate (aq.)
            → silver chloride (s) + sodium nitrate (aq.)
   (b)   Barium nitrate (aq.) + sodium sulphate (aq.)
            → barium sulphate (s) + sodium nitrate (aq.)
   (c)   Lead nitrate (aq.) + potassium chromate (aq.)
            → lead chromate (s) + potassium nitrate (aq.)
   (d)   Barium nitrate (aq.) + potassium chromate (aq.)
            → barium chromate (s) + potassium nitrate (aq.)

Topic D1
1 (a)   Carbon monoxide 28 times the mass of 1 atom of hydrogen
            Nitrogen 28 times the mass of 1 atom of hydrogen
            Nitrogen dioxide 46 times the mass of 1 atom of hydrogen
   (b) –   (c) Sulphur dioxide
2 B
3 D

Topic D2
1 Because carbon is a non-metal. Other non-metals.
2 (a)   Carbon dioxide (g) + water (l) → sugars   Endothermic
   (b)   Sugars → carbon dioxide (g) + water (l)   Exothermic
3 Calcium hydroxide (aq.) + carbon dioxide (g)
            → calcium carbonate (s) + water (l)
   Sodium hydroxide (aq.) + carbon dioxide (g)
            → sodium carbonate (aq.) + water (l)
   Sodium carbonate is soluble
4 –

5   Calcium hydrogencarbonate $\xrightleftharpoons[\text{pass } CO_2]{\text{heat}}$ calcium carbonate + water + carbon dioxide

```
                                  heat
   Calcium hydrogencarbonate ⇌ calcium carbonate + water
                          pass CO₂          + carbon dioxide

          + carbon
          dioxide                heat

   calcium      + water
   hydroxide  ←——————— calcium oxide + carbon dioxide
```

(i)   The cloudiness disappears (calcium hydrogencarbonate forms, and this is soluble)
(ii)   Temporary hard water
(iii)   A scum (calcium stearate) would form.

6   –

Topic D3

1   Six
2   Energy is stored in sugar units. Plants break starch into sugar units and obtain the energy by respiration.
3   –
4   Because so many things are made starting from oil.
5   –
6   Their ability to form chains and rings of carbon atoms.

Topic E1

1   (a)   Sodium carbonate   (b)   Calcium sulphate
    (c)   Potassium nitrate   (d)   Magnesium phosphate
    (e)   Sodium bromide
2   (a)   Nitric acid   (b)   Hydrochloric acid   (c)   Hydroiodic acid   (d)   Sulphuric acid   (e)   Sulphurous acid
3   (a)   Magnesium sulphate (aq.) + hydrogen (g)
    (b)   Zinc nitrate (aq.) + hydrogen (g)
    (c)   Lead nitrate (aq.) + water (l)
    (d)   Potassium chloride (aq.) + water (l)
    (e)   Calcium sulphate (aq.) + carbon dioxide (g) + water (l)
    (f)   Silver chloride (s) + sodium nitrate (aq.)
4   –

Topic E2   (Negative electrode first)

1   Lead; chlorine   2   Hydrogen; oxygen   3   Hydrogen; chlorine
4   Copper; oxygen   5   Copper; chlorine   6   Silver; chlorine
7   Silver; oxygen   8   Hydrogen; oxygen   9   Silver; oxygen
10   Hydrogen; oxygen

INDEX